MATH
100 IDEAS IN 100 WORDS

A whistle-stop tour of key concepts

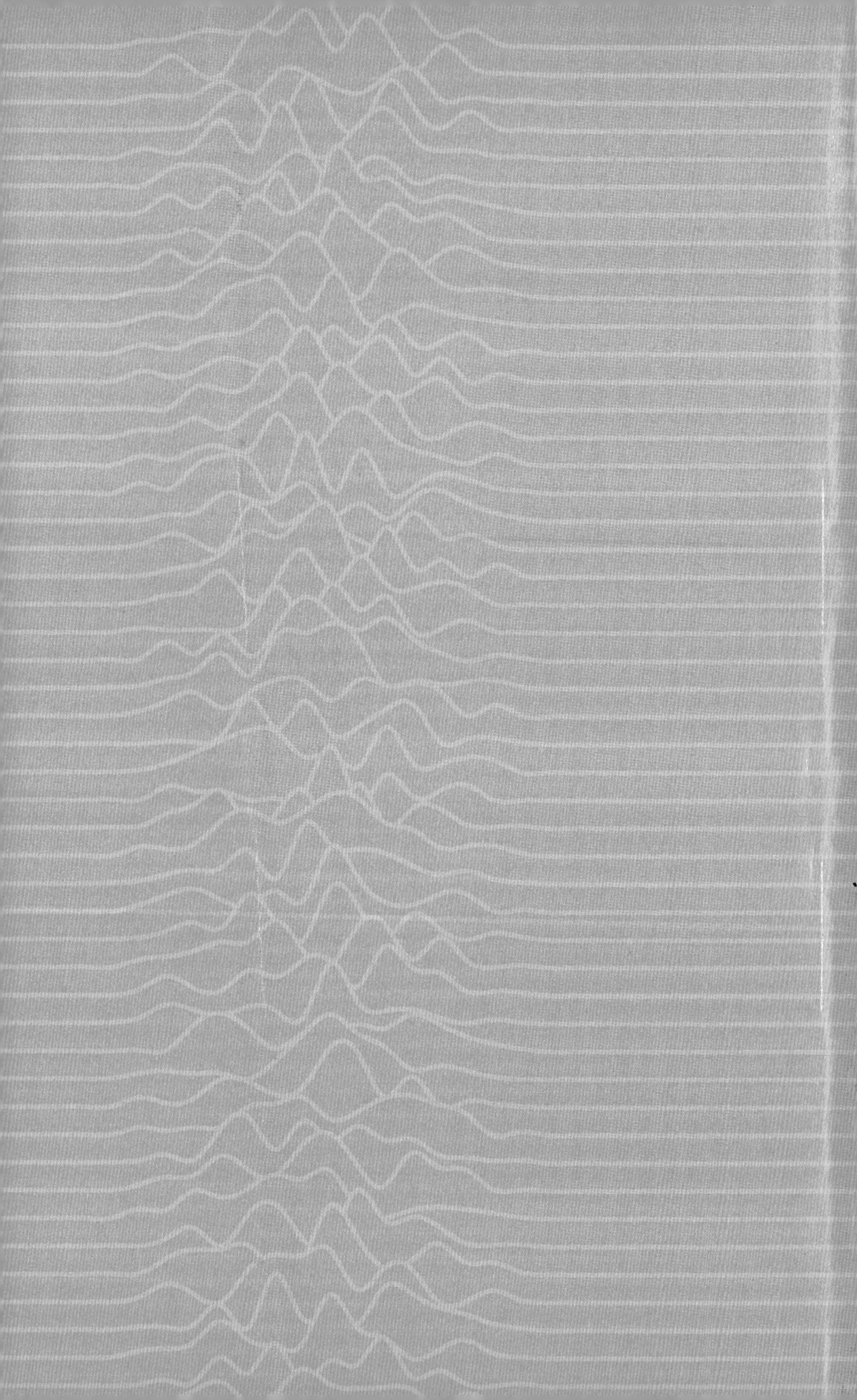

SCIENCE
MUSEUM

MATH
100
IDEAS
IN 100
WORDS

**A whistle-stop tour
of key concepts**

Sam Hartburn, Ben Sparks,
Katie Steckles

Contents

Introduction

Mathematics has existed for as long as the human race has been observing and trying to understand the universe. The first historical uses were for measuring the the physical world, especially in astronomy, and for accounting purposes – in order to work out how much tax someone owed, the area of their fields had to be measured.

From these extremely practical origins, maths has grown into a much broader subject: from simple mensuration (calculating the areas and volumes of shapes) we have developed the field of geometry; from modelling the motion of planets in space we created calculus; from trying to solve simple equations we developed the basic ideas of number theory; and while trying to understand and describe the characteristics of populations we produced the field of statistics.

Early mathematicians were often considered "polymaths", and counted maths as one of their many subjects of interest. They hailed from all over the world: Japan, China, India, Egypt, Greece and Persia (now Iran) were all hugely important historic centres of mathematical thought. Over centuries, ideas were transported around the world by travelling mathematicians, leading to a huge growth in mathematical ideas taking place in Europe from around 1500CE, and eventually developing into modern mathematics – where mathematicians all over the world collaborate and share ideas.

Today, maths has many branches, which employ mathematical thinking for uses way beyond their original conception: as well as working with other subjects to apply mathematics to other sciences, maths is used to model and understand an incredible variety of real-world phenomena, from financial markets to weather forecasting. Starting from the original numerical and geometric approaches, we have created whole new fields of study, including data science, combinatorics and category theory.

Mathematics isn't just useful for its applications in the real world: much of mathematical thinking concerns pure mathematics, which deals with abstract ideas and the fundamental structures that underlie much other mathematics. This includes the maths of maths itself: mathematical logic, and the way we turn ideas into facts by proving theorems.

The future of mathematics is also bright and exciting – not least due to the possibilities opened up by computer technology, including computer proof assistants and the ability to model increasingly complex systems. But there's also huge potential for more collaboration between different fields – both within and outside of mathematics – allowing disparate topics to become more closely connected and make use of the tools others have developed, so that mathematicians become polymaths once again.

Number

For many people, mathematics is synonymous with numbers. Once we've learned to count as children, numbers become an integral part of our lives – we use arithmetic to manage our finances, ratios to adapt recipes in the kitchen and measurements to plan the space around us. But there is more to the mathematics of numbers than these everyday uses.

Some ideas we take for granted now caused controversy when first introduced. The number zero was not accepted in Europe until the fifteenth century, despite being a foundation of both Indian and Arabic mathematics, and imaginary numbers were frowned upon at first by many. Infinity is still a mind-boggling concept full of paradoxes.

And sometimes highly abstract concepts circle back to become useful on an everyday level. Number theory, once considered purely abstract, is now vital to areas such as electronics, computer security and 3D computer graphics. While day-to-day users don't need to understand advanced mathematics, it has given us the tools we need to develop the technology that underpins our daily lives.

1

In 100 words

The counting numbers, also called the natural numbers, are the numbers we count with, starting with 1, 2, 3, ... They are used for counting distinct objects, such as the number of days until the next full moon. The natural numbers are all positive (bigger than zero) – if we also include zero and negative numbers (numbers smaller than zero), then we have the **integers**. Some people use the term "whole number" for the natural numbers, and others use it for all integers; some people include zero in the natural numbers, and others don't. Even mathematicians can't always agree on definitions!

Integers and counting

Evidence of human counting, in the form of notches carved into animal bones, goes back at least 40,000 years. From our earliest development, counting has been part of our culture, long before we had symbols for numbers or formal mathematics.

People have invented many methods to make counting quicker and easier. One example is a chuckrum board: a wooden board with a grid of small indentations used in 18th- and 19th-century India for counting small coins. The coins were spread over the surface until each one was in an indentation, making it easy to count how many there were.

WHY IT MATTERS
Counting is humanity's first experience of mathematics

KEY MOMENTS
The Lebombo bone, a small piece of bone from 44,000 years ago, was discovered in Africa in the 1970s. It is marked with 29 defined notches and is one of the earliest pieces of evidence of human counting

WHAT COMES NEXT
Rational and **real numbers** allow us to quantify values that fall in between integers

SEE ALSO
Rational and real numbers, p.10
Zero, p.13
Integer sequences, p.26

2

Rational and real numbers

Not everything in the world can be described using **integers**. Even simple questions such as, "How many bars of chocolate do I need to make this cake?" might need a non-integer number to answer them. And if you've ever tried to divide a bill between several people, or worked out how much of a loan you have left to pay, you've probably been using **rational numbers** that aren't integers.

Irrational numbers can describe geometrical values that we can't precisely measure. The **square root** of 5 is the length of the longest side of a right-angled triangle whose shorter sides are length 1 and 2 (see page 98), and π is the **ratio** of a circle's circumference to its diameter. This means you can never have a circle whose circumference and diameter are both rational.

Real numbers can be classified as algebraic or **transcendental**. **Algebraic numbers**, which include all rational and some irrational numbers, are all solutions of polynomials (see page 48) with only rational **coefficients**; transcendental numbers are not. Even though we know of very few transcendental numbers – π and *e* are two examples – we do know that there are an **uncountably infinite** number of them.

A rational number can be written as a **ratio** (or fraction) of two integers, such as 2/3 or 99/100. All integers are rational, as any integer is the ratio of itself and 1. As a decimal, a rational number either has a **finite** number of **decimal places** (e.g. 7.99 = 799/100) or the decimal digits repeat (e.g. 0.19191919... = 19/99).

A number whose decimal places go on forever with no repeat is called irrational as it can't be written as a ratio. The irrational numbers include π, e, and the square root of 2.

Together, the rationals and irrationals make up the real numbers.

Irrational

TRANSCENDENTALS
e.g π and e

ALGEBRAIC
IRRATIONALS
e.g $\sqrt{17}$ and Φ

Algebraic

RATIONALS
Including integers

The Real Numbers

3

The real number line

Mathematicians think of the **real numbers** as existing on an infinite line that stretches from large negative numbers, past zero, to large positive numbers in the other direction. This line can be used to visualise something that's quite a difficult concept: an infinite continuum. In an infinite continuum, between any two numbers there's an infinity of other numbers.

We can also think of the real line as a geometrical space – one-dimensional, with distances defined by the distance between two numbers (found by subtracting the smaller from the larger).

WHY IT MATTERS
The space of real numbers is a truly one-dimensional space, which crops up in geometry and topology as well as being an uncountably infinite **set**

KEY THINKERS
John Napier
(1550–1617)
Richard Dedekind
(1831–1916)

WHAT COMES NEXT
Extending the real line to a two-dimensional plane gives us **complex numbers**, which exist in the complex plane

SEE ALSO
Integers and counting, p.9
Rational and real numbers, p.10
The complex plane, p.36

In **100** words

The real numbers can all be placed on an infinite line, which we call the real line. Since numbers have a defined ordering (expressed using "less than" and "greater than"), such a line is well-defined and will stretch off infinitely in both directions.
While a line containing the **integers** would contain a distinct point for each number, the real numbers are dense – this means that given any tiny distance, denoted ε (epsilon), we can find two real numbers less than this distance apart. The real numbers are therefore **uncountably infinite**, as is any **finite** interval of the real numbers.

4

In 100 words

Zero is the whole number that is between the positive numbers and the negative numbers. For most of human history, there has been disagreement about whether zero is a number (and many earlier counting systems simply did not use a zero). We use zero as a placeholder in our base 10 counting system so that we can represent arbitrarily large numbers with only 10 symbols (the digits 0 to 9). However, the system stops making sense if you try to divide by zero. Zero has the special property that when you add it to another number, that number stays unchanged.

Zero

Is nothing a thing? Is it a number? People have long argued over such philosophical questions. However, it is beyond doubt that many things became easier once people became comfortable with the concept of nothing and gave it a name and a symbol.

We use zero as a placeholder to indicate an empty column in "place value" counting systems. For example, 209 is 2 hundreds, 0 tens and 9 ones.

Compare this with the Roman numbering system, which had no zero. This caused problems when writing larger numbers because there was no placeholder to change the value of other symbols.

With zero added, the **real numbers** form a complete number line, but a two-dimensional number line can also be useful (see page 36).

WHY IT MATTERS
Nothing really matters: without the number zero, the operation of addition does not have an **identity** element (see page 78)

KEY THINKERS
Some ancient civilisations developed symbols for zero – for example, the Babylonians (by c. 300 BCE) and the Mayans (by c. 30 BCE). Muhammad ibn Ahmad al-Khwarizmi (c. 9/10th century CE) first suggested a small circle for the number zero

WHAT COMES NEXT
Dividing by zero causes problems in the rest of the number system. Trying to deal with this problem has led to many new ideas (see pages 28 and 58)

SEE ALSO
Equations and inequalities, p.46
Functions, p.54
Group theory, p.76
Statistics, p.142

5

Basic operations

Almost all human activity involving numbers and counting also involves some arithmetic, from measuring and counting money to scoring games and scaling recipes.

Throughout history, humans have used technology to make calculation easier, from early computation devices like the Japanese soroban abacus and Chinese suanpan, through devices like Napier's bones and mechanical calculators, to modern-day computing devices, which can perform a staggering 10^{18} computations per second.

Addition and subtraction are considered **inverse** operations, as are multiplication and division. The basic operations, together with numbers, can be considered as an **algebraic structure** like a **group**.

In
100
words

Arithmetical operations like addition, multiplication, subtraction and division are central to many people's experience of mathematics, and can be thought of formally as simple functions that take two inputs and produce an output.

Multiplication can be thought of as repeated addition – adding something to itself several times is the same as multiplying it by a number. Similarly, the four basic operations can be extended to more complex ideas such as raising a number to a power, taking roots and finding percentages. Parentheses (brackets) can be used to indicate the order in which a larger calculation is intended to be computed.

"Arithmetic has a very great and elevating effect, compelling the soul to reason about abstract number, and if visible or tangible objects are obtruding upon the argument, refusing to be satisfied."

Plato, Greek philosopher

6

Prime numbers

WHY IT MATTERS
Primes are the building blocks of all numbers. Every integer greater than 1 is either prime or a product of primes

KEY THINKERS
Eratosthenes (c. 270BCE)
Euclid (c. 300BCE)
Kamāl al-Dīn al-Fārisī (1265–1318)

WHAT COMES NEXT
Mathematicians have proved the prime number theorem, which describes *approximately* how the primes are distributed, but the exact distribution is still unpredictable. Improving our understanding of how the primes are distributed is bound up with the Riemann hypothesis

Prime numbers, or primes, are **integers** with exactly two **factors**. There are infinitely many of them, but at the time of writing the largest *known* prime (found in 2018) is $2^{82,589,933} - 1$, which has 24,862,048 digits. Primes are the building blocks of all integers – every non-prime (composite) number can be written as a **product** of primes in a unique way.

Factorising numbers (breaking them down into their prime factors) is easy to describe, but for very large numbers it can take a long time. Encryption exploits this fact: the RSA protocol multiplies two large primes together to make a very large number that is released publicly and used to encrypt messages. Anyone wanting to crack the encryption to read a message needs to recover the original two primes. Although this is possible, in practice it takes too long, so the message remains secure.

The number 1 is not a prime number as it has only one factor. If it were prime, then the prime factorisations would not be unique, e.g. $18 = 2 \times 3 \times 3$ and also $1 \times 2 \times 3 \times 3$.

The only even prime is 2, since all other even numbers are divisible by 2.

A prime is a positive whole number with precisely two **factors**: 1 and itself. A factor of an integer is another integer which divides into it exactly, leaving no remainder.

The first prime number is 2. It is followed by 3, 5, 7, 11, 13, 17, 19, 23... These prime numbers continue forever, but not in a regular pattern. Every integer (bigger than 1) can be written as a multiplication involving only primes. This is called the prime factorisation, and the fact that it can be done uniquely for every positive whole number is the fundamental theorem of arithmetic.

There are 25 prime numbers under 100, highlighted in this grid

1	2	3	4	5	6	7	8	9	10
11	12	13	14	15	16	17	18	19	20
21	22	23	24	25	26	27	28	29	30
31	32	33	34	35	36	37	38	39	40
41	42	43	44	45	46	47	48	49	50
51	52	53	54	55	56	57	58	59	60
61	62	63	64	65	66	67	68	69	70
71	72	73	74	75	76	77	78	79	80
81	82	83	84	85	86	87	88	89	90
91	92	93	94	95	96	97	98	99	100

7

The twin prime conjecture

WHY IT MATTERS
Understanding how the primes behave and are distributed is an enduring puzzle for mathematicians

KEY THINKERS
Alphonse de Polignac (1826–1863)
Yitang Zhang (1955–)
The Polymath project (2009–)

WHAT COMES NEXT
If another unproved conjecture (the generalised Elliott–Halberstam conjecture) is true, then the Polymath project's work can be used to prove that the gap (the value of n) is as low as 6

SEE ALSO
Prime numbers, p.16
Infinity, p.27
Proof, p.167

The twin prime conjecture states that there are an infinite number of pairs of prime numbers with a difference of 2 between them. It was originally stated by French mathematician Alphonse de Polignac in 1849 in a different way: "There are infinitely many cases of two consecutive prime numbers with difference n." Some refer to this as the prime gap problem: i.e. the question of what size gaps between primes will keep occurring. In the case $n = 2$, this is the twin prime conjecture.

In 2013, Chinese American mathematician Yitang Zhang proved that a value for n does exist and that its value is less than 70 million. Though it seems at first glance a long way from the twin prime conjecture ($n = 2$), it is still a remarkable improvement on a problem that had seen no progress for a long time. Within a year, a collaborative effort from many mathematicians (including Terence Tao and James Maynard), called the Polymath project, reduced the value of n to 246. The story of the prime gaps and the Polymath project is a striking example of modern collaborative mathematics.

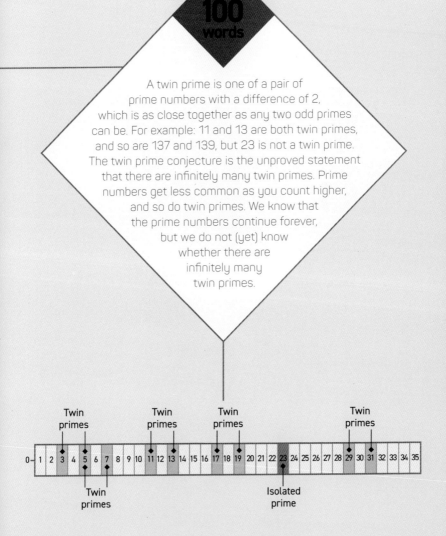

A twin prime is one of a pair of
prime numbers with a difference of 2,
which is as close together as any two odd primes
can be. For example: 11 and 13 are both twin primes,
and so are 137 and 139, but 23 is not a twin prime.
The twin prime conjecture is the unproved statement
that there are infinitely many twin primes. Prime
numbers get less common as you count higher,
and so do twin primes. We know that
the prime numbers continue forever,
but we do not (yet) know
whether there are
infinitely many
twin primes.

Twin
primes
Twin
primes
Twin
primes
Twin
primes

0— 1 2 3 4 5 6 7 8 9 10 11 12 13 14 15 16 17 18 19 20 21 22 23 24 25 26 27 28 29 30 31 32 33 34 35

Twin
primes

Isolated
prime

Goldbach's conjecture

Goldbach's conjecture is one of a number of related conjectures about sums of prime numbers. Perhaps the strongest of these was proposed by Harvey Dubner, and states that every even number greater than 4,208 is the sum of two twin primes, although not necessarily twinned with each other. ("Strongest" in this context means it places the fewest conditions on which numbers it applies to, and therefore covers the most cases.)

If Dubner's conjecture is proven to be true, this will not only prove Goldbach's conjecture but also the twin prime conjecture, as well as a number of weaker conjectures.

In
100
words

Goldbach's conjecture is the claim that every even number above 2 can be written as the sum of two primes. For example, 8 can be written as $3 + 5$, and 24 can be written as $11 + 13$. The conjecture was proposed by Christian Goldbach in 1742 but, despite mathematicians' best efforts, remains unproven. It has been checked by computer for numbers up to 4×10^{18}. A weaker version of Goldbach's conjecture claims that every odd number greater than 5 is the sum of three primes. Harald Helfgott released a widely accepted **proof** of this in 2013, but it has not yet been finally verified.

The **square root** of 2 is approximately 1.4142, because when you multiply 1.4142 by itself you get an answer of approximately 2. The exact value of √2 is impossible to write down in decimal or fraction notation, because it is irrational. Euclid wrote down a **proof** of this fact in his *Elements* (c. 300BCE). To use the value of the square root of 2 in a calculation, mathematicians just leave it written as √2. This is an example of a **surd**. The diagonal of any square is always √2 times the length of its side.

Square root of 2

The **square root** of 2, written as √2, is the number which, when multiplied by itself (i.e. squared), gives you 2. It is an **irrational number** (i.e. not a **ratio** of whole numbers).

In ancient times, the concept of irrational numbers caused problems. Legend has it that an ancient Greek mathematical group called the Pythagoreans, who believed that all numbers were rational, killed one of their own members (Hippasus) because he insisted that √2 was an irrational number.

The ratio of the sides of A4 paper is √2 : 1. Halving any A-size paper gives a smaller rectangle with exactly the same proportions; √2 is the only proportion with this useful property.

WHY IT MATTERS
The square root of 2 was the first number proved to be irrational. Irrational numbers remain very useful, despite being hard to write down; examples include *π* and the golden ratio

KEY THINKERS
Pythagoras and Hippasus
(c. 600–400BCE)
Euclid (c. 300BCE)

WHAT COMES NEXT
Irrational numbers like √2 are **real numbers**, but the square root of –1 is not a real number – it is an **imaginary number** (see page 34)

SEE ALSO
Rational and real numbers, p.10
Logarithms and *e*, p.22
Pi *π*, p.23
The golden ratio, p.24
Proof, p.167

10

Logarithms and e

Logarithms were originally developed to turn difficult multiplications and divisions into much simpler additions and subtractions, using the fact that $log(a \times b) = log(a) + log(b)$ and $log(a \div b) = log(a) - log(b)$. John Napier wrote the first table of logarithms in 1614. Later, slide rules were developed for calculating logarithms; these were faster and more convenient than looking up values in a table.

The constant e is ubiquitous in mathematics, and has many applications. The function e^x can be used to model natural phenomena such as bacterial growth and radioactive decay, as well as man-made constructs such as continuous compound interest.

WHY IT MATTERS
Logarithms and e have wide and varied applications, including in finance, statistics and psychology

KEY THINKERS
John Napier
(1550–1617)
Leonhard Euler
(1707–1783)

WHAT COMES NEXT
Both $ln(x)$ and e^x are important functions in differential and integral calculus, and e is related to the trigonometric functions via Euler's formula

SEE ALSO

In
100
words

A logarithm tells us what **power** we need to raise a base number to in order to get another number. For example, 10 raised to the power of 3 is 1,000, so the base 10 logarithm of 1,000 is 3. This is written mathematically as $log_{10}(1000) = 3$. The most commonly used base is Euler's number, e, an **irrational number** that starts 2.71828.... Base e logarithms, written $log_e(x)$ or $ln(x)$, are called natural logarithms, and the function e^x is called the natural exponential function. Both $ln(x)$ and e^x **diverge** to infinity, but while e^x does so very quickly, $ln(x)$ diverges very slowly.

If you divide the circumference (the distance around the edge) of any circle by its diameter (the distance across the middle, through the centre), you'll always get the same answer: a value slightly over 3. The Greek letter π is used to represent this value. π also relates the area of a circle to its radius (the distance from the centre to the edge); the area is equal to π multiplied by the square of the radius. We'll never know the exact value of π, because it's an **irrational number**, but as a decimal, its first ten digits are 3.141592653...

Pi (π)

The value of π, pronounced "pi", is used in many fields, including architecture, engineering, astronomy and statistics. For most practical purposes an accuracy of two **decimal places**, 3.14, is good enough to calculate measurements to millimetre accuracy. NASA uses 15 or 16 digits in calculations for its navigational and positioning systems, and the Committee on Data for Science and Technology uses 32 digits to compute values of fundamental physical constants. But mathematicians can't resist the challenge of computing many more digits of π. In 2022, Emma Haruka Iwao used advanced computing techniques to calculate 100 trillion digits.

> **"Love is like pi – natural, irrational, and very important."**
> **Lisa Hoffman**, entrepreneur

WHY IT MATTERS
π is used in many practical calculations, including calculating the pitch of a guitar string or the load a column can carry before buckling

KEY THINKERS
The concept of π has been known since ancient times. Archimedes (third century BCE) devised a method to approximate the value of π using polygons. Leonhard Euler (1707–1783) popularised use of the symbol π

WHAT COMES NEXT
An open question is: are the digits of π normal? This means, is any sequence of **integers**, of any length, equally likely to appear within them?

SEE ALSO
Rational and real numbers, p.10

12

The golden ratio

The golden **ratio**, or Φ, is a number that shows up in many areas of mathematics. Perhaps the most striking is the number of places it appears in a regular pentagon and the pentagram that can be inscribed within it. In each of the triangles indicated in the diagram, dividing the length of the longer edge by the length of the shorter one will give Φ.

The number Φ is irrational, and it can be considered the *most* **irrational number** as it is the hardest to approximate with a fraction. One way to find an approximate value for Φ is using the Fibonacci sequence. This sequence starts 1, 1, and further terms are generated by adding the two previous terms, so the sequence goes 1, 1, 2, 3, 5, 8, 13, 21, 34, ... You can find the ratio between two consecutive terms by dividing the larger by the smaller: $1 \div 1 = 1$, $2 \div 1 = 2$, and carrying on gives 1.5, 1.666..., 1.6, 1.625, 1.615..., 1.619... These ratios get closer and closer to the value of $\Phi = 1.61803...$

The golden ratio is thought to be aesthetically pleasing, with architects such as Le Corbusier using it extensively in their designs.

Choose any two numbers; call the larger one a and the smaller one b. Use them to calculate two new numbers $a \div b$ and $(a + b) \div a$. Are your two answers close to each other? If they are, then the numbers are close to being in the golden **ratio**. Commonly denoted by the Greek letter phi (Φ), the golden ratio is an irrational number, equal to $(1+\sqrt{5})/2$, or approximately 1.61803....

The first known definition of phi appears in Euclid's *Elements*. Euclid called it an "extreme and mean ratio", and it has gained a reputation for being aesthetically pleasing.

In each of these triangles, dividing the length of the longer edge by the length of the shorter one will give Φ

Integer sequences

For as long as numbers have been studied, they have been grouped into sequences and given names; for example, a prime p is called a Sophie Germain prime if $2p + 1$ is also prime. The Online Encyclopaedia of Integer Sequences collects examples in an online database for mathematicians to reference.

Patterns in numbers can connect unexpected areas of mathematics. For example, John McKay noticed that a particular **set** of numbers found in the study of **modular functions** was also associated with the **monster group**. This was called the monstrous moonshine discovery and led to collaboration between the fields.

WHY IT MATTERS
The mathematical study of patterns and the relationships between numbers underpins many other areas of mathematics

KEY THINKERS
Fibonacci (1170–1250)
Sophie Germain (1776–1831)
Neil Sloane (1939–)
John McKay (1939–2022)

WHAT COMES NEXT
The $(n+1)$th term in the sequence

SEE ALSO

In
100
words

Numbers can be considered as part of **finite** or infinite sequences in which they share a property or are defined by a rule. For example, the prime numbers have no easily predictable pattern but have a shared property, and the Fibonacci numbers are defined by a rule relating each term to the previous two. For the sequence of square numbers, the nth term is produced by squaring the nth whole number.
Integer sequences often turn up as the solution to problems in number theory, or help to describe patterns in sets of mathematical structures, such as in geometry or algebra.

14

Infinite means to continue forever, without bounds. There are infinitely many counting numbers – however big a number is, you can always add one to it (to get a new, bigger number). Things can also be infinitely small. Dividing the number 1 by larger and larger numbers will give an answer that gets closer and closer to zero, but it will never be equal to zero – the gap between the answer and zero becomes infinitely small. And the seemingly paradoxical nature of infinity means that however small that gap is, it will always contain an infinite number of **real numbers**.

WHY IT MATTERS
Thinking about infinity helps us to make sense of the seemingly infinite universe we live in

KEY THINKERS
Ancient Greeks
(1200–323BCE)
Jain mathematicians
(India, c. 400BCE)
Bernard Bolzano
(1781–1648)
Georg Cantor
(1845–1918)

WHAT COMES NEXT
Calculus is built on the idea of infinitesimals, or infinitely small values. The study of limits looks at what happens to infinite sums of numbers

SEE ALSO

Infinity

Mathematicians and philosophers have been grappling with the concept of infinity since ancient times. The ancient Greeks were divided into two schools of thought: those who believed that matter could continually be divided into smaller and smaller pieces, and those (the Atomists) who believed that you would eventually reach an indivisible particle. In India, Jain texts from around 400BCE describe a recursive process for generating huge numbers, but state that however many times this process is carried out, it will still not reach the highest possible number.

Georg Cantor (1845–1918) made a distinction between different sizes of infinity. The natural numbers are **countably infinite**, as is any **set** of numbers that can be paired one-to-one with the natural numbers, such as the set of square numbers. The real numbers cannot be paired one-to-one in this way. However you define the pairs, there will always be a real number that hasn't been paired. They are **uncountably infinite**.

15

Limits

WHY IT MATTERS
Many fundamental
concepts in maths rely
on the notion of a limit,
including calculus,
continuous functions,
and analysis

KEY THINKERS
Bernard Bolzano
(1781–1848)
Augustin-Louis Cauchy
(1789–1857)
G. H. Hardy
(1877–1947)

WHAT COMES NEXT
A continuous function
is one with no sudden
jumps or gaps. We can
use limits to formulate
a precise definition of
continuity

SEE ALSO
Infinity, p.27
Functions, p.54
Calculus, p.58

Limits are important when studying the behaviour of functions, and are part of the definition of the **derivative** and **integral** in calculus. Because infinity can behave in strange ways, limits need to be handled carefully.

We can define the precise conditions when a sequence has a limit, using a notional small number denoted by epsilon (ε) – the limit exists if for any given ε you can find a point along the sequence beyond which all the terms are closer to the limit than ε. It's like a game – if you pick a small value of ε, such as 0.1, 0.001 or 0.000001, I can tell you how far to go along the sequence to get within that distance of the limit.

A constant sequence like 1,1,1,1... has a limit of 1, but for a sequence like 1,2,1,2... the limit does not exist, since there's no one single value the numbers are getting closer to.

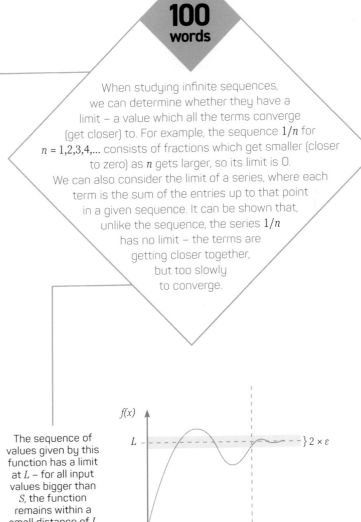

When studying infinite sequences, we can determine whether they have a limit – a value which all the terms converge (get closer) to. For example, the sequence $1/n$ for $n = 1,2,3,4,...$ consists of fractions which get smaller (closer to zero) as n gets larger, so its limit is 0. We can also consider the limit of a series, where each term is the sum of the entries up to that point in a given sequence. It can be shown that, unlike the sequence, the series $1/n$ has no limit – the terms are getting closer together, but too slowly to converge.

The sequence of values given by this function has a limit at L – for all input values bigger than S, the function remains within a small distance of L

$f(x)$

L

$\}\, 2 \times \varepsilon$

S

x

16

Number bases

The base of a number system is the number of unique symbols used to represent digits. Base 10 is natural to humans because we have 10 fingers, but different bases can be used for different purposes. For example, in computer systems, digits are represented by the current in an electrical circuit being on or off – there are only two options, so base 2 is ideal. Base 2 uses only two symbols, 0 and 1; each digit is multiplied by a **power of 2**. You need a much longer number in base 2 than in base 10 to represent the same value – the base 10 number 874 is 1101101010 in base 2.

These long strings of digits can be difficult to read, but it's computationally intensive to convert between base 10 and base 2, so computer programs often use base 16: because 16 is a power of 2, it's quick for computers to convert between these two bases. Numbers in base 16 use the symbols 0 to 9 along with the letters A to F and are short enough to be easily readable by humans – the decimal number 874 is 36A in base 16, which means A (or 10) ones, 6 sixteens, and 3 two-hundred-and-fifty-sixes.

"There are 10 types of people in the world. Those who understand binary, and those who don't."

Famous mathematical joke

When we write numbers using our standard counting system, we use the symbols 0 to 9; there are ten symbols, and we say the number is written in base 10 (decimal). To write numbers higher than 9 we use more digits, and each digit is multiplied by a power of 10, starting from $10^0 = 1$ in the far right column. The power increases as you move to the left, giving us tens, hundreds, thousands, and so on. So the number 874 means 4 ones, 7 tens and 8 hundreds. Other common number bases are base 2 (binary) and base 16 (hexadecimal).

17

Fermat's last theorem

Fermat's last theorem concerns a problem in number theory, and achieved fame among mathematicians partly because of the tantalising way in which it was written – Fermat scribbled it in the margin of a book, along with a note saying that he had a marvellous way to prove it, but not enough space to write it out. These scribblings were published 30 years later, after his death.

Over the next 300 years, mathematicians including Sophie Germain proved it for specific values of n, but an overall **proof** seemed impossible. Andrew Wiles took a completely different approach, using modern techniques that Fermat couldn't have known about. Whether or not Fermat really did have a proof remains a mystery!

WHY IT MATTERS
Until it was proved, this was one of the most famous and long-standing open questions in mathematics

KEY THINKERS
Pierre de Fermat (1607–1665)
Sophie Germain (1776–1831)
Andrew Wiles (1953–)

WHAT COMES NEXT
Andrew Wiles' work on proving Fermat's last theorem led to the development of new techniques in number theory which have been used to prove many further theorems

SEE ALSO
Proof, p.167
Pythagoras' theorem, p.98

In 100 words

We know from Pythagoras' theorem that there are infinitely many **integer** solutions to the equation $a^2 + b^2 = c^2$; for example, $3^2 + 4^2 = 5^2$. Fermat's last theorem states that if n is greater than 2, there are no integers a, b, and c for which $a^n + b^n = c^n$. This is an example of a **Diophantine equation** (see page 46). The theorem was conjectured by Fermat in approximately 1637, but remained unproven until Andrew Wiles published a valid proof in 1995. The proof initially came from an observation about seemingly unrelated mathematical objects known as **elliptic curves**, and the two areas turned out to be related in surprising ways.

In 100 words

Modular arithmetic is a system of counting using a **finite** set of numbers that wraps around to zero after reaching a certain value, called the **modulus**. It's sometimes called clock arithmetic, since a clock face has numbers 1 to 12, and so its arithmetic is performed modulo 12: if we add 3 hours to 11, we get 2. A large number can be considered modulo a smaller number by finding the remainder on division, or residue. This allows us to categorise all numbers given a modulus; for example, 7, 12, and 142 all have a value of 2 modulo 5.

WHY IT MATTERS
Modular arithmetic has very many uses, including in understanding musical scales, in computer algebra, in public key cryptography and in the design of technology

KEY THINKERS
Carl Friedrich Gauss (1777–1855)
Ralph Merkle, Whitfield Diffie, Martin Hellman (public key cryptography, 1976)

WHAT COMES NEXT
Numbers modulo n form finite **groups**

SEE ALSO

Modular arithmetic

Modular arithmetic is based on ideas described by Gauss in the book *Disquisitiones Arithmeticae* (1801). It uses only a small finite **set** of numbers, and arithmetic is performed within that set only. It can be used for simplifying calculations and as an underlying framework for other mathematical structures. For example, matrices can be constructed with all entries and calculations taken modulo n.

Check digits on barcodes allow a barcode scanner to check it's read a number correctly, by checking that the answer to a calculation using the barcode digits matches the final digit when considered modulo 10.

19

Imaginary numbers

You can't take the **square root** of a negative number. Or rather, you can't do it if you want your answer to be a **real number**.

If, however, you accept that there might be a type of number that is somehow different to the real numbers, then you can define a new number i where $i^2 = -1$. This number i is called the imaginary unit or number, and sometimes described as "the square root of negative one".

Adding **imaginary numbers** to real numbers defines what mathematicians call **complex numbers**, and they turn out to be spectacularly useful.

Complex numbers make many other mathematical facts simpler and easier to connect: for example, the fundamental theorem of algebra (see page 48), and exponentials and trigonometry (see page 72).

The first glimpses of imaginary numbers becoming useful occurred in the 16th century while Italian mathematicians were seeking new methods of solving cubic equations (see page 48).

An imaginary number is a multiple of i, where $i^2 = -1$. Using these numbers lets us extend the real numbers to the complex numbers. René Descartes, in the 17th century, didn't like the idea and coined the name "imaginary" as an insult, but the name has stuck, and although most people don't encounter them at school level, complex numbers rank among the most useful and important ideas in all of mathematics. The contrast between their apparent abstractness (partly due to the unhelpful label "imaginary") and their usefulness in electronics, telecommunications, and many other areas is something many people find remarkable.

"To simplify, you should 'complexify'. That is, when you have a complicated problem and wish to simplify it, it is a good idea to replace all reals by complex numbers."
Benoit Mandelbrot, quoting **Gaston Julia**

20

The complex plane

WHY IT MATTERS
Without complex numbers, most of our modern technological world wouldn't exist. Our understanding of electricity, flight, differential equations, and quantum physics (among many other things) relies on them

KEY THINKERS
Leonhard Euler (1707–1783)
Abraham de Moivre (1667–1754)
Carl Freidrich Gauss (1777–1855)
Jean-Robert Argand (1768–1822)
Caspar Wessel (1745–1818)
Nikolay Zhukovsky (1847–1921)

WHAT COMES NEXT
Complex numbers connect exponentials and trigonometry

SEE ALSO
Rational and real numbers, p.10
Imaginary numbers, p.34
Quaternions, p.38
Polynomials, p.48
Differential equations, p.64
Euler's formula and identity, p.72

If you add a **real number** to an **imaginary number**, the result is called a **complex number** (the word complex originally meant a thing made up of more than one part, like a "gym complex", for example).

To visualise how these work, we can update our image of numbers from a one-dimensional line (see page 12) to a two-dimensional **plane** (the complex plane), where every point is a number consisting of a real part and an imaginary part.

Complex numbers allow us to solve differential equations that model the real world, from simple oscillations to the equations used in hospital MRI scanning machines.

They also help us develop aircraft and wing shapes. Engineers analyse airflow around a simple cylinder shape and then use complex numbers to transform this simple shape and the airflow into the best shape to generate lift (an aerofoil).

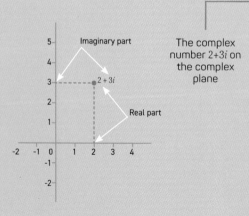

The complex number $2+3i$ on the complex plane

Complex numbers are two-dimensional numbers which are the sum of real and imaginary numbers.

The complex plane is a two-dimensional number grid which helps us visualise complex numbers. Any point on the plane represents a complex number. It acts very much like a coordinate grid, but the x coordinate is the real part of the number and the y coordinate is the imaginary part.

When complex numbers are represented on the complex plane in this way, ideas of shape and space (geometry) become obviously connected to ideas of number (arithmetic). This has generated new insights into many areas of mathematics.

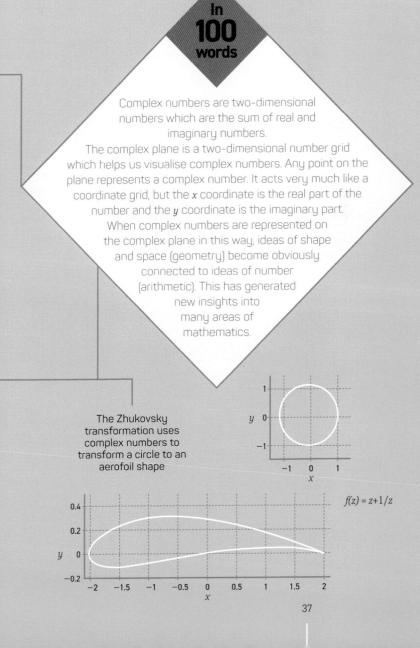

The Zhukovsky transformation uses complex numbers to transform a circle to an aerofoil shape

$$f(z) = z + 1/z$$

37

21 Quaternions

Complex numbers let us use multiplication to describe rotations in two dimensions: multiplying a complex number plotted as a point on the complex **plane** by $cos(45°) + sin(45°)\,i$ rotates it by 45° anticlockwise about the **origin**. Quaternions extend the complex numbers in a way that lets us use a similar technique to compute rotations and detect orientation in three dimensions. This is how a smartphone knows which way up it is, how robots are able to move in three-dimensional space, and how aerospace navigation systems work.

WHY IT MATTERS
As well as being important in many areas of pure mathematics, quaternions provide an efficient way to calculate movement and have widespread uses in computer graphics and navigational software

KEY THINKERS
William Rowan Hamilton (1805–1865)

WHAT COMES NEXT
Octonions extend the quaternions into eight dimensions, having a real part and seven imaginary parts

SEE ALSO
Imaginary numbers, p.34
The complex plane, p.36

In **100** words

Quaternions extend the complex numbers into four dimensions. Complex numbers have a real part (a **real number**) and an imaginary part (a real number multiplied by the imaginary unit, i). Quaternions extend this to have one real part and three imaginary parts. This is achieved by introducing two more imaginary units, j and k, with the property that $i^2 = j^2 = k^2 = i \times j \times k = -1$. So the quaternion $3 + 5i + 2j + 7k$ has a real part of 3 and imaginary parts $5i$, $2j$, and $7k$. A multiplication table defines how to multiply these imaginary units; for example, $i \times j = k$ but $j \times i = -k$.

In **100** words

When an exact value for a quantity cannot be stated – either because it's a real-world quantity that has been imprecisely measured, or a number with an infinite decimal expansion – we can write it as an approximation instead. The "approximately equals" symbol $\approx$, introduced by Alfred Greenhill, is used when an **expression** is not exact: we can write $\pi \approx 3.14$.

Approximations can be given to a number of **decimal places** (two, in the example above), or a number of significant figures – non-zero values either side of the decimal point. 0.000345, 436,000, and 1.23 are all approximations to three significant figures.

Approximation

It's often impossible to measure and record precise numerical values, especially for real-world quantities like length and temperature, where measurement can only be made to a certain degree of accuracy. It's also impossible to write down an exact value for numbers like π or e, whose decimal expansion is infinite and never repeats.

Approximations let us write numbers concisely when less precision is needed. This depends on what they are used for. For example, astronomers deal with huge numbers, and might give numbers as a multiple of a power of 10 – one astronomical unit is approximately 1.496×10^8 km, or around 150 million km.

WHY IT MATTERS
Storing data to unnecessary levels of precision wastes space and makes calculations take longer – but understanding the context is important

KEY THINKERS
Alfred Greenhill (1847–1927)
Brook Taylor (1685–1731)

WHAT COMES NEXT
Approximations can also apply to functions and shapes – for example, we can use Taylor series to approximate functions using simpler ones

SEE ALSO
Taylor series, p.50
Fourier series, p.74

23

Percentages

Since ancient Rome, calculations have been made which use multiples of 1/100 of a quantity, and percentages (from the Latin, *per centum* meaning by a hundred) are still widely used today. Percentages allow us to easily understand proportions of a whole and are used to describe and compute increases and decreases to quantities.

Percentages are also crucial to expressing probabilities, since they are values between zero and one. They are widely used in business and finance to describe interest rates, profit, and discounts, in science for describing compositions of mixtures and experimental error, and for slopes of roads.

WHY IT MATTERS
Percentages are a versatile way to express probabilities, proportions, rates, and relative differences across a wide range of applications

KEY POINT
It's important not to confuse decimals and percentages – 0.4 is equivalent to 40%, but 0.4% is a hundred times smaller

WHAT COMES NEXT
Diagrams like pie and waffle charts can be used to visualise percentages, and box plots show error bars representing percentage error

SEE ALSO
Rational and real numbers, p.10
Probability, p.141
Statistical diagrams, p.144

In
100
words

Percentages are used to express quantities or proportions as a value out of 100. 100% represents the whole of a quantity, the fraction ½ corresponds to 50%, and decimal numbers less than 1 can be converted to percentages using multiplication by 100: the decimal number 0.27 corresponds to 27%. A percentage is a **dimensionless quantity**.

Expressing values as percentages means proportions can be easily compared across sets of different sizes. Percentages can also be used to express changes in value – an increase from 30 to 45 would be a 50% increase, or we can say 45 is 150% of 30.

Combinatorics is the study of combinations and permutations: ways of combining and ordering selections of objects. Given a **set** of objects, combinatorics has formulae for calculating the number of different **subsets** of a given size, i.e. the number of ways to choose x objects from n. Other formulae can be used to calculate the number of ways to rearrange a set of objects into a different order. These types of calculations are useful in many areas, and can be used and combined to count possibilities, measure the complexity of computer algorithms, and determine the many possible outcomes of probabilistic experiments.

WHY IT MATTERS
Combinatorics is related to virtually every branch of mathematics and drives discoveries in geometry, computing, and optimisation theory

KEY THINKERS
Sushruta
(India, c. 500BCE)
Plutarch (Greece, c. 46–119CE)
Rabbi Abraham ibn Ezra (c. 1090–1165)

WHAT COMES NEXT
A small increase in the number of possibilities can lead to a huge growth in combinations, known as a combinatorial explosion

SEE ALSO
Integers and counting, p.9
Graph theory, p.126

Combinatorics

Combinatorics allows us to answer questions about the number of possible arrangements of **finite** sets of objects – for example, how many different poker straight flushes are there in a deck of 52 cards?

The related area of design theory also lets us work out whether structures of multiple **subsets** satisfying given criteria exist – from the design of an experiment (allowing participants in a trial to be assigned different treatments in all possible combinations) to a series of golf games (allowing all players to play each other while avoiding one particular pairing due to a falling-out).

25

Units of measurement

WHY IT MATTERS
Standardised units of
measurement allow
people to be confident
that they are talking
about the same value

KEY MOMENTS
The Metre Convention,
a diplomatic treaty that
forms the basis of
international
agreement on units of
measurement, was
originally signed by 17
countries in 1875, and
now has 51 member
states

WHAT COMES NEXT
New prefixes *ronna* and
quetta (for octillions
and nonillions) and
ronto and *quecto* (for
octillionths and
nonillionths) were
introduced in 2022

SEE ALSO
Integers and
counting, p.9

Traditionally, units were based on objects that were readily available, and often specific to a local area. Even today, UK shoe sizes are based on the size of a barleycorn, now standardised to one third of an inch – each size is defined as how many barleycorns longer or shorter it is than a defined base size of 12, so a size 11 is one barleycorn smaller than a size 12.

Before the seventeenth century there were no international standard units for weight. Merchants trading internationally relied on conversion tables, which were often inaccurate. In 1818, the UK government collected copies of standard weights for all countries where they had an overseas consul. These were stored at the Royal Mint and used to create accurate conversion tables for merchants.

The International System of Units (SI) defines the units officially recognised by most countries. They were originally based on physical artefacts stored in a vault, but since 2019, all SI units have been redefined in terms of known physical constants. For example, one metre, once based on the length of a physical stick, is now defined by a calculation using the speed of light and the resonant frequency of the element caesium.

A unit of measurement is a defined quantity that other quantities can be compared to. The SI base units are second (time), metre (length), kilogram (mass), ampere (electric current), kelvin (thermodynamic temperature), mole (amount of substance), and candela (luminous intensity). Base units are modified by **prefixes** for convenience at different scales – we use picometres for atomic lengths, and gigametres for astronomical distances. There are 22 named units derived from the base units. For example, newtons, used to measure force, are derived from kilograms, metres, and seconds. Other derived units are unnamed: speed, for example, is measured in metres per second.

"When you can measure what you are speaking about, and express it in numbers, you know something about it; but when you cannot measure it, when you cannot express it in numbers, your knowledge is of a meagre and unsatisfactory kind."
Lord Kelvin

Algebra

The basics of algebra will be familiar to anyone who's ever been asked to "find x". We use letters to represent unknowns, and find their values by solving equations. With variables like this we can make general statements, and define functions: given a value x, we can find $2x$, or x^2, and study how the output varies when we change the input.

This is all fundamental to many mathematical ideas, particularly calculus: the study of change. Whether it's a moving object, a waveform, or the motion of a fluid like water, we can describe and study change by modelling things with functions and observing their behaviour.

We also use algebra to describe points in space: coordinates provide a link to spatial geometry. Simultaneous equations allow us to combine the definitions of multiple relationships, describe complicated systems and optimise for particular criteria. Algebra also includes the study of abstract structures such as groups, which describe objects interacting, from addition of numbers to twists of a puzzle cube.

But generally, algebra in its many forms is only about one thing: abstraction. Instead of concrete real-world structures, algebra allows us to create notional worlds where we can play, tweaking numbers to understand how things work generally rather than in specific cases.

Algebra is the process of using symbols instead of numbers to analyse and manipulate relationships. It allows us to **generalise** results from specific cases to wider ones using formulae. For example, the formula speed = distance ÷ time (or $s = d \div t$) is an algebraic formula that tells us how speed, distance and time are related. If we know two of these values, the formula lets us calculate the third. Algebra is used to show complicated relationships between many variables in fields such as **multivariate analysis** in statistics. It is also used for more abstract concepts such as group theory.

WHY IT MATTERS
Instead of giving lots of specific examples, algebra lets us give one general formula

KEY THINKERS
There is evidence that the ancient Babylonians used algebra in their calculations. The ninth-century Muslim mathematician Muhammad ibn Musa Al-Khwarizmi is known as the inventor of algebra

WHAT COMES NEXT
The results of algebraic analysis are used in modelling and forecasting to predict future events and trends

SEE ALSO

Algebra

Algebra can help you keep track of a number even if you don't know what it is. You may have tried a "think of a number" magic trick: ask somebody to choose any number, double it, add six, halve the result, and subtract the original number. Then astound them by revealing that you know their answer is three!

Applying algebra makes it clear why. If you start with x, double it and add 6, you'll get $2x + 6$. Halving this gives $x + 3$, and subtracting x leaves 3, no matter what value x has. Algebra helps reveal the pattern explaining the mystery.

27

Equations and inequalities

Equations and inequalities are statements that compare two values or **expressions**. Equality is an important concept in mathematics, and it can have different meanings in different contexts. Numerical quantities are equal if their value is the same, whereas two functions (see page 54) might be considered equal if they produce the same output for every input.

Coordinate **axes** can be used to visualise both equations and inequalities: for example, plotting all the points where $x^2 + y^2 = 1$ generates a circle; $x^2 + y^2 < 1$ is the region inside the circle, and $x^2 + y^2 > 1$ is the region outside it.

Some equations can be solved to find the values of unknown quantities; others have an infinite number of solutions, or none.

Attempts to solve equations can lead to brand new areas of mathematics. **Imaginary numbers** (see page 34) were invented as a solution to the equation $x^2 = -1$, which has no real solution.

Often, mathematicians are interested in whether or not a set of equations can be solved, without necessarily knowing what the solutions are. For example, a **Diophantine equation** is one in which only **integer** solutions are valid; it was proved in 1970 that there is no algorithm to decide if such an equation has any solutions.

> **"So far, all of reality seems to be described by exquisite, elegant mathematical equations. We can't stop now – it's got to be beautiful all the way down!"**
>
> **Katie Mack**, physicist

An equation is a statement containing an equals symbol with an **expression** on each side. Expressions can include numbers ($1 + 2 = 3$), letters ($3x = 6y$), and other mathematical symbols; what's important is that the two values on each side of the equals symbol are equal. Inequalities are similar to equations, but tell us that one side is greater or less than the other. There are four inequality symbols: $<$ (less than), $\leq$ (less than or equal to), $>$ (greater than), and $\geq$ (greater than or equal to). Other symbols used to compare expressions include $\approx$ (approximately equal to) and $\neq$ (not equal to).

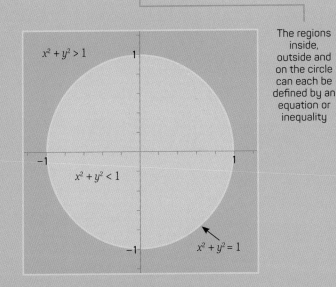

$x^2 + y^2 > 1$

$x^2 + y^2 < 1$

$x^2 + y^2 = 1$

The regions inside, outside and on the circle can each be defined by an equation or inequality

28

Polynomials

WHY IT MATTERS
Polynomial equations describe many real-world situations. For example, the path of a falling object can be modelled with a parabola, a degree 2 polynomial

KEY THINKERS
Brahmagupta
(c 600CE)
Niels Henrik Abel
(1802–1829)
Évariste Galois
(1811–1832)

WHAT COMES NEXT
Polynomials can be used to approximate other functions in a way that helps to understand their properties or behaviour (see page 50)

SEE ALSO
The complex plane, p.36
Equations and inequalities, p.46
Taylor series, p.50
Conic sections, p.102

A polynomial is an algebraic **expression** that is a sum of positive **integer** powers of a variable, such as $x^2 - 5x + 5$ or $x^3 - x^2 - 5x + 2$.

The largest power determines the degree of the polynomial, which gives us information on its properties and shape when graphed. For example, $x^2 - 5x + 5$ is a degree 2 polynomial, known as a quadratic. When graphed, a quadratic expression always gives a **parabola** shape. A first-degree polynomial is called linear, because its graph is a straight line.

Solving polynomial equations has been important since ancient times. Quadratic equations written in the form $ax^2 + bx + c = 0$ can be solved using the formula

$$x = \frac{-b \pm \sqrt{b^2 - 4ac}}{2a}$$

The formula finds the coordinates of the points where the parabola crosses the x-axis.

Methods for solving cubic and quartic equations also exist, but there are no simple algebraic methods for solving quintic or higher degree polynomial equations. In the 18th and 19th centuries, people sometimes used physical and mechanical devices to solve them, but these are now largely replaced by computer techniques.

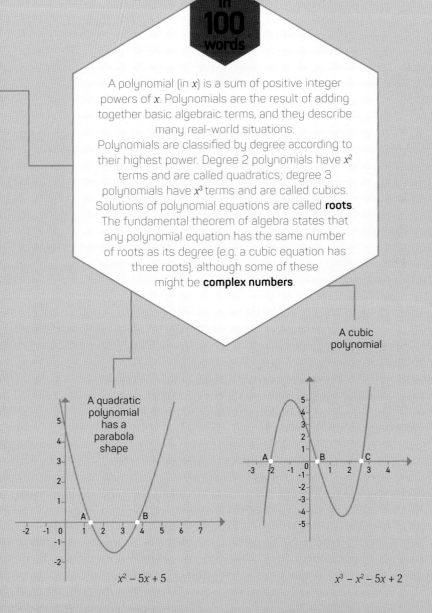

A polynomial (in x) is a sum of positive integer powers of x. Polynomials are the result of adding together basic algebraic terms, and they describe many real-world situations.

Polynomials are classified by degree according to their highest power. Degree 2 polynomials have x^2 terms and are called quadratics; degree 3 polynomials have x^3 terms and are called cubics. Solutions of polynomial equations are called **roots**. The fundamental theorem of algebra states that any polynomial equation has the same number of roots as its degree (e.g. a cubic equation has three roots), although some of these might be **complex numbers**.

A cubic polynomial

A quadratic polynomial has a parabola shape

$x^2 - 5x + 5$

$x^3 - x^2 - 5x + 2$

29

Taylor series

A Taylor series is a way of rewriting a function as a polynomial (see page 48). This is often useful to understand its properties or approximate its behaviour. For example, the function $sin(x)$ can be approximated by a cubic polynomial: $x - x^3/6$.

It's a good approximation, but not perfect for the whole function. Increasing the degree of the polynomial used will improve the approximation. Using an infinite number of terms gives an *exact* match of the original function. This infinite polynomial is called the Taylor series. For example:

$$sin(x) = x - \frac{x^3}{6} + \frac{x^5}{120} - \frac{x^7}{5040} + \ldots$$

Cutting off (truncating) a Taylor series at some point gives a Taylor polynomial, and these are useful approximations for a function. Truncating the $sin(x)$ series after only one term gives an approximation $sin(x) \approx x$, which is widely used for small angles.

Another important example of Taylor series is for e^x (see page 22): $e^x = 1 + x + x^2/2 + x^3/6 + \ldots$

If you differentiate the Taylor series you will notice the important property of the e^x function: it is its own **derivative** (and is the only function with this property).

The function in green is approximated with increasing numbers of terms of its Taylor Series (red, blue, then orange). The horizontal bars show the increasing size of the region where the approximation is quite good

A Taylor series is a (possibly infinite) polynomial which can represent another function – for example, *sin(x)*. Writing functions as polynomials, even infinite ones, can reveal properties about the functions that might not be obvious. The polynomials can also provide helpful approximations. A full Taylor series for a non-polynomial function will involve an infinite number of terms, but it can be stopped somewhere (truncated) to give Taylor polynomials. These provide the approximations which help simplify formulas when modelling otherwise unsolvable problems. Higher degrees provide better approximations. The Taylor series of a function might be considered a definition of the function itself.

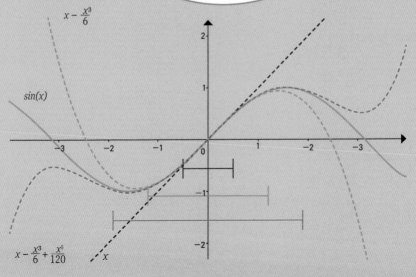

30

Simultaneous equations

When a situation involves multiple interacting relationships between unknown variables, simultaneous equations allow us to capture all of this information and try to find a solution. If there are more unknown variables than equations, we won't be able to obtain a unique solution, but we will likely be able to constrain the problem to a smaller set.

Systems of linear equations are often described using matrices (see page 66). Since each equation describes a line or plane in n-dimensional space, the solution can be thought of as the point where these intersect, so linear algebra has deep links to geometry.

An equation describes a relationship between variables, e.g. $y = 2x$. If we're describing a more complex system, we can use two or more equations that are all true at the same time – called simultaneous equations. Given n unknowns and n different equations describing how they're related, the system can often be uniquely solved to find values which satisfy all the constraints.

Most real-world situations need more than one variable, so simultaneous equations are used to model them. Examples include financial relationships and physical systems. The equations can be linear (a sum of the variables), nonlinear (involving powers), or even differential equations.

Functions

WHY IT MATTERS
Functions are crucial to almost all fields of mathematical thinking. They can be used to model real-world behaviour, and they form the basis of calculus

Mathematical functions describe mappings from one **set** to another, and express how a quantity varies depending on another quantity – for example, the function which maps x to $x + 5$ will give outputs that increase as x increases. Functions can be applied to any set – numbers, abstract objects, or shapes – or even a combination of sets, as long as the output can be defined.

KEY THINKERS
Nikolai Lobachevsky (1792–1856)
Peter Gustav Lejeune Dirichlet (1805–1859)
Augustin-Louis Cauchy (1789–1857)

Functions are conventionally denoted f, and we write $f(x)$ to describe the result of applying f to an input x. For functions like $f(x) = x^2$, where the input and output are both a single number, we can plot the function on a graph, with input values on the horizontal axis and the corresponding values of $f(x)$ on the vertical axis.

WHAT COMES NEXT
Mathematicians study the properties of functions: whether they can be inverted, how they behave as the input approaches a particular value (see page 28), and whether they can be differentiated (see page 60)

For functions where the input is a point in two-dimensional space, or a **complex number**, this isn't possible, and we need to use other methods to visualise the function.

SEE ALSO
Prime numbers, p.16
Limits, p.28
Calculus, p.58

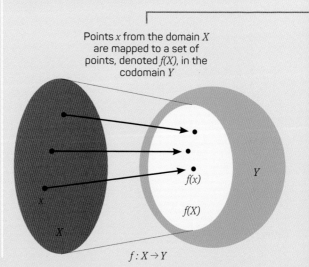

Points x from the domain X are mapped to a set of points, denoted $f(X)$, in the codomain Y

Y

$f(x)$

$f(X)$

x

X

$f : X \rightarrow Y$

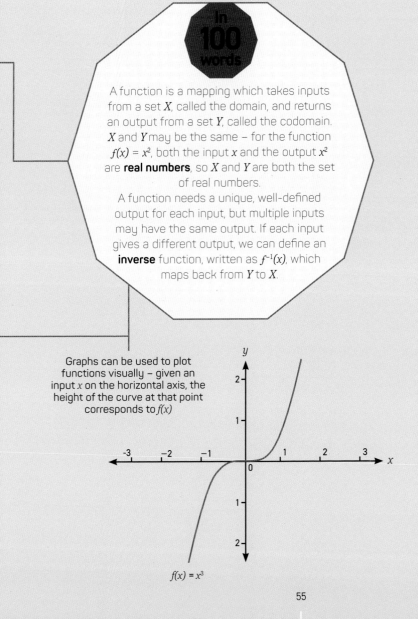

A function is a mapping which takes inputs from a set X, called the domain, and returns an output from a set Y, called the codomain. X and Y may be the same – for the function $f(x) = x^2$, both the input x and the output x^2 are **real numbers**, so X and Y are both the set of real numbers.

A function needs a unique, well-defined output for each input, but multiple inputs may have the same output. If each input gives a different output, we can define an **inverse** function, written as $f^{-1}(x)$, which maps back from Y to X.

Graphs can be used to plot functions visually – given an input x on the horizontal axis, the height of the curve at that point corresponds to $f(x)$

$f(x) = x^3$

32

The Riemann hypothesis

One of the Millennium Prize Problems, the Riemann hypothesis concerns the behaviour of a particular function called the Riemann zeta function, and has connections to various other mathematical questions.

The function can be restated as a sum over the prime numbers, so there's a connection to number theory – including implications for the arrangement and distribution of prime numbers, as well as patterns in prime factorisations of numbers and square numbers via the Möbius function. The zeta function also has connections to applied statistics and quantum field theory.

Mathematicians are often concerned with the input values to a function which give zero as an output – called the zeros of the function – and in this case, assuming the hypothesis holds, there's a pattern to be found.

"If I were to awaken after having slept for a thousand years, my first question would be: Has the Riemann hypothesis been proven?"

David Hilbert, physicist

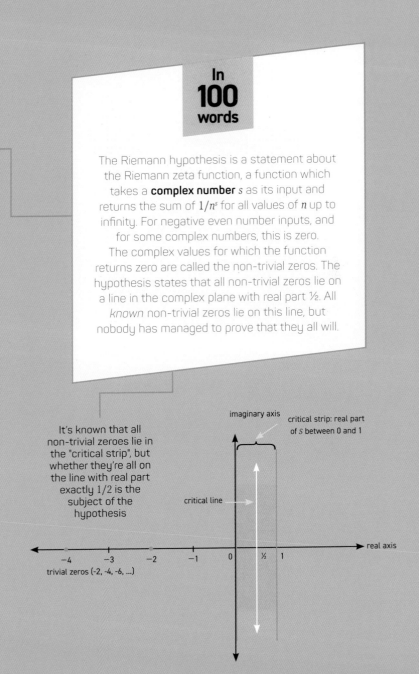

The Riemann hypothesis is a statement about the Riemann zeta function, a function which takes a **complex number** s as its input and returns the sum of $1/n^s$ for all values of n up to infinity. For negative even number inputs, and for some complex numbers, this is zero. The complex values for which the function returns zero are called the non-trivial zeros. The hypothesis states that all non-trivial zeros lie on a line in the complex plane with real part ½. All *known* non-trivial zeros lie on this line, but nobody has managed to prove that they all will.

It's known that all non-trivial zeroes lie in the "critical strip", but whether they're all on the line with real part exactly 1/2 is the subject of the hypothesis

imaginary axis

critical strip: real part of s between 0 and 1

critical line

real axis

−4   −3   −2   −1   0   ½   1

trivial zeros (-2, -4, -6, ...)

Calculus

Imagine a ball that's just been thrown. Freeze an instantaneous moment in time and you can capture the ball's position, but you then have no information about how far the ball has travelled, or how quickly it's moving, or how its movement will change in the next moment. In the 17th and 18th centuries, Isaac Newton and Gottfried Leibniz independently developed ways of dealing mathematically with infinitely small moments of time called infinitesimals (see page 27) in order to do useful analysis of things that change.

Calculus (which originally just meant "calculation") is short for "the calculus of infinitesimals".

Both branches of calculus – differentiation and integration – work by examining the limit of situations where you break things into smaller and smaller pieces.

Historically there was significant controversy around which of Newton or Leibniz could claim to be the first to describe the methods of calculus, although the generally accepted consensus now is that they appear to have developed their ideas independently, arriving at similar results by different methods and notations.

> "The calculus is the greatest aid we have to the application of physical truth in the broadest sense of the word."
>
> **William Fogg Osgood**, mathematician

In 100 words

Calculus is the study of anything that changes or moves. If you can represent a change or movement as a mathematical function, then you can ask useful questions about the *way* that it changes. The two main questions in calculus involve the steepness or **gradient** of the function (i.e. how quickly it changes) and the area under the function (i.e. how things build up over time). Calculation of the gradient uses a process called differentiation, and calculation of the area involves a process called integration. The study of these two concepts has proved spectacularly useful in modelling the real world.

34

Differentiation

Differentiation, part of calculus, is the process of finding the **rate of change** (or the **derivative**) of a function, which is equivalent to the steepness or **gradient** of the graph of the function.

You can differentiate further to find the rates of change of the rates of change. For example, given a function describing the position of an object, the first derivative is the rate of change of the position, which is the **velocity** ("how quickly does position change?"), and the second derivative is the rate of change of the velocity, which is the acceleration ("how quickly does velocity change?").

Knowing about the gradient can help determine important points on a function. For example, at maximum or minimum points, the gradient will be flat (i.e. have a zero slope). This process can be used to find optimum strategies, such as the most efficient or profitable choice.

There is more than one notation for differentiation. If a function is written as $y = f(x)$, its derivative might be written as $\frac{dy}{dx}$ (pronounced "dee y by dee x" – Leibniz's notation), but also sometimes as $f'(x)$ (pronounced "f dash of x" – Newton's notation) to emphasise that it is a new function.

Differentiation is the process of finding the gradient or slope of a function. This is equivalent to finding a new function that describes the **rate of change** of the original function. The new function is called the gradient function, or derivative, and the process can be repeated if necessary to find further derivatives. Calculating the derivatives of a function provides useful information about the function, in particular where turning points (e.g. maximum and minimum points) occur. We can also work backwards: if we have information about the derivatives of a function, we can reverse the process to find the original function.

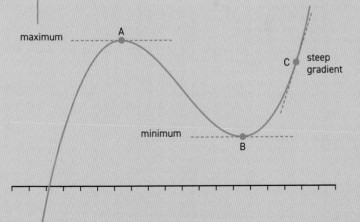

The maximum and minimum points of a function can be found using differentiation

maximum — — — — A — — — — — —

C ● steep gradient

minimum — — — — B — — — —

35

Integration

Integration, part of calculus, is the process of finding an area under the graph of a function (finding the **integral**). This is equivalent to finding how the function accumulates over time. This is useful in many situations. For example, if you know a function for an object's velocity (giving a velocity–time graph) then integrating it will give you a function for how far the object has travelled (its displacement).

Integration can be thought of as the limit of a sum of lots of small areas, so techniques from integration are useful in many applications that need a sum of many small pieces. Examples include deriving the standard formulae for areas and volumes of shapes, finding centres of mass of physical objects, and finding the expected average outcome from a **probability distribution**.

The techniques of integration can be generalised to find uses in much wider areas when combined with other ideas such as **complex numbers** and trigonometry (for example, Fourier Transforms – see page 74).

Many real-world systems are modelled using differential equations and solved using integration (although many different techniques are needed).

Integration is the process of determining how a function accumulates, equivalent to finding an area under the function. It is the **inverse** process to differentiation, and together they form the basis of calculus.

The area under a curve can be approximated by drawing many small rectangles and adding their areas up. The smaller the rectangles, the better the approximation, and using this method to define integration is called Riemann integration.

The relationship between differentiation and integration (and using integration to calculate an area) is known as the **fundamental theorem of calculus**. Integration techniques are used in solving many real-world problems.

The integral of a function represents an area between the graph and the x-axis

Area under the function between 2 points

A   B

36

Differential equations

Differential equations combine algebraic relationships with ideas from calculus, creating equations involving derivatives (see page 60) which can be used to model many real-world systems, including motion in space, heat transfer and fluid flow.

Like any other kind of equation, differential equations can also be combined into systems of equations describing a problem, and solutions to the systems can be found by integration, or by numerical methods, which involve approximating curves by straight line sections. This is often the only option, as many systems can be shown to not be directly solvable.

In
100
words

A differential equation is a
relationship between variables which
involves a **derivative**. The order of a differential
equation is determined by the highest level of
derivative involved, so second-order equations involve
second derivatives, which describe the **rate of change** of the rate
of change. Differential equations can often be expressed as
polynomials in the derivative of one or more variables.
Partial differential equations (PDEs) are used when
a function involves multiple variables, but you
want the rate of change with respect to
only one of them. If the function
involves only one variable,
they are called ordinary
differential equations
(ODEs).

Matrices

Mathematicians have long studied grids of numbers, called matrices, and used them for solving simultaneous equations and for linear programming (see page 132). They were later used in describing geometric transformations like rotations and reflections.

Since the development of computers, matrices have become hugely important because they enable efficient large calculations. Software like MATLAB was developed to work with matrices; combining and manipulating matrices allows computers to calculate, handle 3D graphics, and solve **optimisation problems**. The speed at which a computer can process calculations involving matrices has often been used as a benchmark for processing power.

A matrix is a two-dimensional grid or array. Matrices can be used to describe – among other things – the positions of objects in space, collections of constraints like systems of equations, steps in an iterative process, or transformations of shapes. They can have any number of rows and columns, and the entries can be numbers, symbols, or mathematical **expressions**.

Matrices can be combined using matrix addition – by adding each pair of entries together – and matrix multiplication, which may result in a matrix of a different size than those being multiplied, with the result depending on the order of multiplication.

$$\begin{bmatrix} 1 & 2 & 3 \\ 4 & 5 & 6 \end{bmatrix} \times \begin{bmatrix} 7 & 8 \\ 9 & 10 \\ 11 & 12 \end{bmatrix} = \begin{bmatrix} 58 & 64 \\ 139 & 154 \end{bmatrix}$$

Matrix multiplication involves taking the entries from one row of the first matrix, multiplying them by one column of the second matrix, and adding them together to make one entry in the product

Permutations

In many situations, we need to consider the ways to reorder a **set** of objects – from rearranging a word into an anagram to designing a medical experiment and randomising the samples. We can describe these shuffles using permutations, which can be thought of as elements of a **group**.

Combinatorics (see page 41) involves counting the numbers of permutations and combinations. The number of possible permutations for a set of n elements is given by $n!$ (n factorial), the product of n with all the numbers smaller than it, for example, $4! = 4 \times 3 \times 2 \times 1 = 24$.

"It is easy to perceive that the prodigious variety which appears both in the works of nature and in the actions of men, and which constitutes the greatest part of the beauty of the universe, is owing to the multitude of different ways in which its several parts are mixed with, or placed near, each other."

Jakob Bernoulli, mathematician

A permutation is a reordering of a set of n elements, and the set of permutations for a given n forms a group, called a symmetric group, which is denoted S_n. Permutations can be thought of as a function from the set to itself. They can be written as a list of the elements of the set and their images, or more compactly as a set of cycles – listing the subsets that form closed loops.

Permutations can be combined using the operation of applying one, then the other, and any permutation can be rewritten as a combination of two-element swaps.

This permutation represents the way the first rotor on the German Railway Enigma machine transforms the alphabet – shown as a list and cycles

$$\begin{pmatrix} ABCDEFGHIJKLMNOPQRSTUVWXYZ \\ JGDQOXUSCAMI\,FRVTPNEWK\,BL\,Z\,YH \end{pmatrix} = (AJ)(BGUKMFXZHSEOV)(CDQPTWLI)(NR)$$

39

Coordinate systems

WHY IT MATTERS
Coordinate systems have many applications, including in navigation, 3D graphics, space travel, and astronomy

KEY THINKERS
René Descartes
(1596–1650)

WHAT COMES NEXT
Coordinate systems with more than three dimensions can be used to model higher-dimensional physical space or more abstract spaces, like data sets

SEE ALSO
Algebra, p.45
Equations and inequalities, p.46
Trigonometry, p.94

A coordinate system is a way to turn a position into numerical values – it creates a bridge between geometry and algebra. It gives you the ability to "see" what an equation looks like: the equation $y = x + 1$, for example, can be translated to "all the places where the vertical coordinate is 1 higher than the horizontal coordinate", which lie on a straight line. Anything from a simple circle to a 3D representation of a person dancing can be turned into an equation and manipulated algebraically; this is the basis of how computer-generated graphics work.

The ability to precisely communicate a location or direction is vital for navigation. Because the Earth is spherical, the lines of latitude and longitude on standard maps are part of an **angular coordinate system**. Latitude is the angle from the **plane** that goes through the equator, and longitude is the angle from a perpendicular plane through the Greenwich meridian. Before satellite GPS systems became commonplace, navigators at sea used devices such as sextants to measure the angles between celestial objects, which allowed them to calculate the current time at Greenwich. By comparing this with local time, they could determine their angle from Greenwich, and thus their longitude.

Coordinate systems allow us to use numbers to precisely define a location. The most familiar is the Cartesian system, which uses numbers to indicate distances along horizontal and vertical **axes**. It is used in maps throughout the world. It can be extended to describe three-dimensional space by adding a third axis, **perpendicular** to the other two. It can even be extended into higher dimensions by adding further axes, although these don't represent the physical world. Polar coordinates in two dimensions are composed of an angle around from the horizontal axis and a distance from the centre point of the axis.

Any point on the globe can be defined by its angle from Greenwich combined with its angle from the equator

Angle from the equator

Angle from Greenwich

40

Euler's formula and identity

WHY IT MATTERS
The formula links apparently separate areas of mathematics together and lets us use complex numbers, which are potentially very abstract, to describe real-world systems in a concrete way

KEY MOMENT
Euler first demonstrated the formula by comparing the Taylor series of e^x, $cos(x)$, and $sin(x)$ and noticing what happened when $x = i\theta$ is substituted into the e^x series

WHAT COMES NEXT
Representing complex numbers as exponentials allows modelling of all sorts of real-world phenomena such as the alternating current produced in electricity generation

SEE ALSO
Logarithms and e, p.22
Imaginary numbers, p.34
The complex plane, p.36
Taylor series, p.50
Trigonometry, p.94

Any point on the unit circle can be represented by $(cos(\theta), sin(\theta))$. If this is treated as a **complex number** on the complex plane, it is $cos(\theta) + i\,sin(\theta)$. The circle can be scaled up (by r) to give any complex number on the plane.

Euler's formula lets us turn this into an exponential formula, which can be used to represent any complex number in the form $re^{i\theta}$. This viewpoint proves fundamental to understanding many of the properties and applications of complex numbers.

If the value of x in the formula is π, you get $e^{i\pi} = -1$, sometimes rewritten as $e^{i\pi} + 1 = 0$. This is called Euler's identity, and it is often claimed to be the most beautiful result in mathematics. The juxtaposition of five fundamental numbers (e, i, π, 1, and 0) in a simple result is remarkable, but Euler's formula itself, relating **exponentials** and trigonometry, is what causes the profound connections.

Euler's formula also explains why the solutions to some second-order differential equations (see page 64), which typically involve exponential functions, can have trigonometric terms present – for example, when describing oscillating behaviour observed in the real world (like a bouncing spring in vehicle suspension).

The imaginary unit $\sqrt{-1}$

$$e^{ix} = cos(x) + i\,sin(x)$$

Euler's number: e

The trigonometric functions cosine & sine

Euler's formula relates exponential functions to the trigonometric functions by the use of **imaginary numbers**. It joins together several areas of mathematics and allows for multiple representations of important and useful concepts.

Euler's formula is $e^{ix} = cos(x) + i\,sin(x)$, where e is Euler's number, i is the imaginary unit, $cos(x)$ and $sin(x)$ are the circular trigonometric functions cosine and sine, and x is any real value. The formula gives us two different ways of representing any complex number: with trigonometry (on the right-hand side) and with **exponentials** (on the left-hand side).

Any point on the unit circle can be thought of as a complex number $cos(\theta) + i\,sin(\theta)$

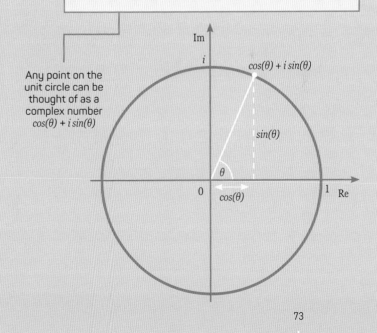

41

Fourier series

Waves of various shapes, sizes and speeds are all around us. Water waves, sound waves – anything that vibrates or wobbles. The simplest wave shape is a sine curve (see page 94), but most waves in the real world are more complicated shapes – the sound of a note from a violin gives a sawtooth-shaped wave.

In the 19th century, Joseph Fourier suggested that any repeating wave shape can be made by adding together multiple sine waves of different frequencies and heights. This has proved extremely useful in understanding waves, and the method of representing a function as the sum of simple trigonometric functions gives us a Fourier series.

Fourier series can be used with **complex numbers** to create what is known as a Fourier Transform, which has revolutionised the way we understand signal processing in the modern world. Methods of doing this mathematics quickly have become known as Fast Fourier Transforms - described as the most important mathematical algorithm of our lifetime.

> **"The Fast Fourier Transform: the most important numerical algorithm of our lifetime."**
>
> **Gilbert Strang**, mathematician

Given any periodic function, which repeats the same shape over and over, a Fourier series is a way of expressing that function as a sum of sine curves of different magnitudes and frequencies. This is how synthesisers create artificial but realistic sounds – mimicking real sounds by overlapping digital sounds.

In reverse, it means we can break down a given real-life wave into simple components in order to analyse it and understand its properties. For example, given a complicated sound wave like someone singing, we can determine the frequencies present in the sound and understand why it sounds like it does.

The blue sawtooth wave can be made by adding all the white sine waves together

42

Group theory

The structure of a **group** – some objects, together with a way to combine them – occurs throughout mathematics. As well as infinite examples like the **integers**, groups can be finite, like a **set** of whole numbers under the operation of modular arithmetic.

Sets of permutations (see page 68) form groups, as do the symmetries of a geometric shape. Another example is the Rubik's cube, where moves on the cube, combined by performing one move after another, form a finite group with over 43 billion billion elements. These examples have different combining operations but share an underlying group structure.

	0	1	2	3
0	0	1	2	3
1	1	2	3	0
2	2	3	0	1
3	3	0	1	2

Cayley tables like this one can be used to describe the group operation – here, it's addition modulo 4, so 2 + 3 = 1 (the remainder of 5 on division by 4)

In 100 words

Groups are mathematical structures consisting of a set of objects which can be combined using a **group operation**. Many familiar structures are groups: for example, integers combined using addition.

There are a few properties that a set and its operation must have to form a group. There must be a unique element in the group which functions as an **identity**: when combined with other elements, it has no effect. Each element must also have an **inverse**: an element combined with its inverse gives the identity. For integers under addition, the identity is zero and the inverse of a number n is $-n$.

43

The identity element

The **identity** element of a **group** or other **algebraic structure** is in many ways the most important element: it provides an anchor for the other elements of the group, and it forms part of the definition of the **group operation**. The **axioms** of groups state that in order for something to be a group, it must have an identity – and the nature of the identity depends on the group operation and structure.

WHY IT MATTERS
Even though the identity is in some sense the element that does nothing, it's an important part of the definition of the structure

KEY THINKERS
Leonhard Euler
(1707–1783)
Niels Henrik Abel
(1802–1829)
Évariste Galois
(1811–1832)
Arthur Cayley
(1821–1895)
Sophus Lie
(1842–1899)

WHAT COMES NEXT
Once a group structure is defined we can use known tools from group theory to study it

SEE ALSO
Zero, p.13
Matrices, p.66
Group theory, p.76

In 100 words

The identity element is the unique element with the property that combining it with any other element under the **group operation** returns that same element. For multiplication of numbers the identity element is 1, and for addition it is 0. The identity permutation is the one which doesn't move anything, and the identity matrix has 1s along the diagonal and 0s everywhere else. The identity element is also what you get back if you combine any group element with its **inverse**. For example, adding a number to its corresponding negative number gives 0, and multiplying n by $1/n$ gives 1.

Geometry

One crucial aspect of mathematics often doesn't involve
numbers at all: the study of shape, and the symmetries of
shapes and patterns, forms the important mathematical
field of geometry. As well as 2D and 3D shapes, it includes
higher-dimensional objects and infinite tiling patterns.

Topology is a more abstract way to study shapes,
considering only their fundamental properties, the
number of holes. We can apply it to real physical shapes,
but also to abstract shapes like data sets, to identify the
shapes made by trends and correlations.

Geometry also links to functions – conic sections describe
shapes made by slicing through a cone, but also link to
trigonometry. Vectors describe motion in space, and have
their own rules for combining and adding. We can also
consider different ways to measure the distance between
two points, and think about how 3D shapes curve and how
this affects the behaviour of shapes on their surface.

Geometry remains a good example of how mathematics
can be both concrete and abstract: shape and space will
always feel more tangible than many mathematical ideas,
but the concepts developed from them can still be
surprisingly abstract.

44

Polygons

WHY IT MATTERS
Giving names to the shapes we see around us lets us understand their properties and uses

KEY THINKERS
Ancient Greeks (as early as 7th century BCE) including Euclid (c. 300BCE)

WHAT COMES NEXT
In three dimensions, solid shapes are called polyhedra. Equivalents in four and higher dimensions exist and are called polytopes, but they are harder to visualise

SEE ALSO
Polyhedra, p.82
Tilings and tessellations, p.84
Circle, p.104

A two-dimensional shape made up of only straight edges is called a polygon. Polygons are classified according to their number of sides (or angles), normally using a Greek word for the number, followed by *gon*. Three-sided and four-sided polygons are exceptions: a triangle arguably should be a trigon (see page 94), and a four-sided shape is a quadrilateral but arguably should be a tetragon. After four sides, the naming becomes more consistent: five – pentagon; six – hexagon; seven – heptagon; and so on.

A circle is not a polygon (as it has no straight edges), but a regular polygon with a very large number of sides is almost indistinguishable from a circle.

Further distinctions between polygons can be made based on their properties. A **convex** polygon is one with no indentations, or where all corners point outwards. For example, a parallelogram is a convex polygon, but a dart (or arrowhead) is not. This definition is equivalent to saying that if you join any two points inside the shape with a line, then that line is also entirely inside the shape. This definition can be generalised to other mathematical objects like **sets** and functions (see page 54).

In 100 words

Polygons are two-dimensional shapes made up of only straight edges. The word comes from the Greek words *poly*, meaning "many", and *gon* ("corner"/"angle"). Examples of polygons include familiar shapes like triangles, squares, rectangles, pentagons (five-sided polygons), and hexagons (six-sided polygons). When all edges of a polygon are the same length (equilateral) and meet at the same angle (equiangular), it is called a regular polygon. Polygons can have many other properties, such as being convex, **concave**, **simple**, or **cyclic**. Some polygons, e.g. squares and regular hexagons, make good shapes for tiles because they tessellate the plane, while others, e.g. pentagons, don't.

Some examples of regular polygons, starting with an equilateral triangle and going up to a regular dodecagon

A parallelogram is convex, but a dart is not

45

Polyhedra

Polyhedra (singular: polyhedron) are three-dimensional shapes (solids) with flat faces which are all polygons.

For example, a cube is a polyhedron with six square faces, but a cylinder is not a polyhedron because it has a curved face.

There are many classes of polyhedra, including Platonic solids (see summary opposite); Archimedean solids, where all faces are regular and the corners are identical, but there is more than one type of face – for example, a truncated icosahedron; and Johnson solids, where the faces are regular but can be of different types, and there is more than one type of corner – for example, a square-based pyramid.

Many structures in the natural world use these shapes at a microscopic level. For example, the structure of *Braarudosphaera bigelowii* – a type of phytoplanktonic algae – is that of a dodecahedron.

WHY IT MATTERS
We live in a three-dimensional world. Naming and understanding the three-dimensional shapes we are surrounded by helps us understand and describe our physical world

KEY THINKERS
Plato (c. 428-348BCE)
Archimedes
(c. 287–212BCE)
Johannes Kepler
(1571–1630CE)

WHAT COMES NEXT
In two dimensions there are infinitely many regular polyhedra (polygons); in three dimensions there are only five – the Platonic solids. In four dimensions there are only six, and in every dimension after that there are only three

SEE ALSO
Polygons, p.80

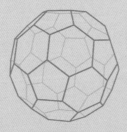

A truncated icosahedron

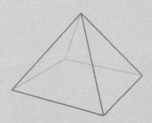

A square based pyramid with equilateral triangle sides

A polyhedron is a three-dimensional shape, or solid, with flat polygonal faces. Polyhedra are the simplest solids to define because they consist only of straight edges and flat faces.

There are only five regular polyhedra – where all faces are regular and identical, meet in the same way at the corners, and are **convex**, i.e. there are no indents. These are the tetrahedron, with 4 triangular faces; the cube, with 6 square faces; the octahedron, with 8 triangular faces; the dodecahedron, with 12 pentagonal faces; and the icosahedron, with 20 triangular faces. These five polyhedra are called the Platonic solids.

The 5 Platonic solids:
1. tetrahedron, 2. cube,
3. octahedron,
4. dodecahedron, and
5. icosahedron

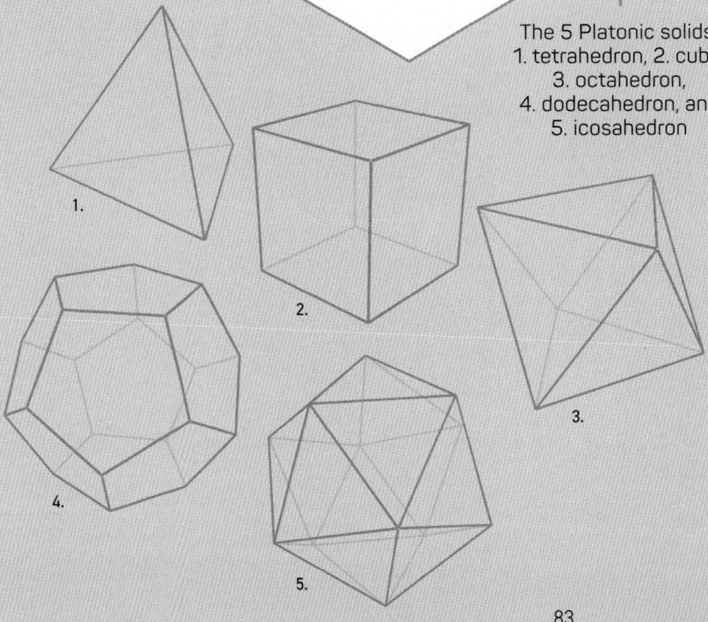

1.

2.

3.

4.

5.

46

Tilings and tessellations

A tessellation, or tiling, is a way of covering a surface using shapes that fit together without overlapping, and without leaving any gaps. A periodic tiling produces a pattern that repeats – you can slide the tiling across or down until it exactly overlays the original. A nonperiodic tiling never repeats, even if it is extended over an infinite surface, and an aperiodic tile set is a set of tiles that can only produce a nonperiodic tiling.

The first aperiodic tile set was found by Robert Berger in 1962, and consisted of 20,426 different shapes. In 1974, Roger Penrose found a set of two shapes, a kite and a dart, that can tile nonperiodically. Penrose's tiles also tile periodically if arranged differently, so they do not form a true aperiodic tile set – but by adding rules about how the tiles may be placed, aperiodicity can be forced. A design based on the tiles was embossed on a quilted toilet paper – until Penrose's development company served a writ against the manufacturer.

In 2023, a group of four mathematicians found two different aperiodic monotiles (single aperiodic tiles) – the "hat" and the "turtle". One of the discoverers, Craig Kaplan, declared on Twitter that he would "absolutely welcome Aperiodic Monotile toilet paper".

Some people were unsatisfied, because some tiles need to be reflected (flipped over) for the tiling to work. Just a few months later, though, the same researchers revealed the "spectre" – an aperiodic monotile that doesn't need reflections.

A regular tiling is made from identical regular polygons. There are three regular polygons that can tessellate: equilateral triangle, square, and hexagon. A monohedral tiling uses only one type of tile. Although a regular pentagon cannot tessellate by itself, there are 15 types of pentagon that form monohedral tilings, the most recent of which was discovered in 2015. A semiregular tiling is made with two or more regular polygons, and has the same configuration at each vertex (intersection of three or more tiles). There are eight possible semiregular tilings, each using two or three different regular shapes.

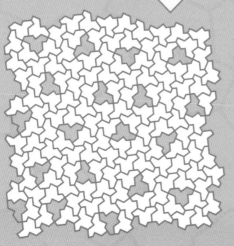

The hat monotile, from Smith, Myers, Kaplan and Strauss' 2023 paper. The shaded tiles are flipped-over versions of the unshaded tiles

47

Duality

By creating a map from one structure to another, duality allows us to look at ideas from different points of view. Although duality is used in many areas of mathematics, one of the easiest ways to understand it is through geometry.

A polyhedron is a three-dimensional shape consisting of vertices, edges, and faces. You can find the dual of a polyhedron by changing each face into a vertex, and joining two vertices from neighbouring faces with an edge. The dual of each Platonic solid (see page 85) is another Platonic solid – a cube and an octahedron form a dual pair, as do an icosahedron and a dodecahedron. The dual of a tetrahedron is a tetrahedron – it is self-dual.

You can construct the dual of a tessellation in a similar way: the centre of each polygon becomes a vertex, and vertices on neighbouring polygons are joined by edges. Combinations of tessellations and their duals are used in many of the intricate tiling structures seen in Islamic architecture, and although they are often highly decorated, the underlying geometric structure is still apparent.

The dual of a cube is an octahedron, and vice versa

Duality is a way of mapping one concept or structure to another, while preserving particular properties. This means that theorems that are proved for one situation can also be proved for its dual. For example, a polyhedron and its dual have the same symmetries, so a theorem about the vertices of a polyhedron could be reframed to apply to the faces. Techniques such as Fourier transformation allow us to find duals in many situations, including for functions, **sets**, and **groups**. Often a dual is easier to work with, and we can apply any conclusions back to the original form.

The dual of a
tiling pattern

48

Symmetry

WHY IT MATTERS
Symmetry and asymmetry are key parts of our understanding of beauty, not just in the visual world but also in music, poetry, and mathematics. The question of invariance: "when does something change and when does it stay the same?" has profound connections throughout mathematics and physics

KEY THINKERS
Euclid (c. 300 BCE)
Emmy Noether
(1882–1935)

WHAT COMES NEXT
Developing notation for symmetries, and how transformations either change or don't change objects, leads to group theory

Visually pleasing objects are sometimes described as symmetrical, and often this is due to parts matching or being in proportion. A butterfly might be considered beautiful not just because of its colouring but also because of its symmetry (the two wings are reflections of each other).

Mathematicians use the word more precisely to mean a way in which a thing looks the same, or acts in the same way (even after it's been changed).

Visually, reflection, rotation, translation (shifting), and scaling are all mathematical examples of transformations. If the object looks the same after being transformed it is said to have a symmetry.

A butterfly is a good example of reflective symmetry

Snowflakes commonly have 6-fold rotational symmetry under a microscope, as well as many lines of reflective symmetry, due to the hexagonal structure of ice molecules

Symmetry in mathematics is the property of an object staying unchanged (invariant) after being transformed in some way. It can be found in visual or physical structures, as well as more abstract mathematical objects like functions, equations, logical connections, and **groups**. The opposite concept is asymmetry, where an object is somehow different after being transformed.

For example, a square has four lines of reflectional symmetry, but also four different rotations that leave it looking the same. A circle is even more symmetrical – it looks the same after being reflected in any diameter or rotated through any amount about its centre.

Reflectional symmetry and asymmetry

Symmetric

Asymmetric

49

Topology

Topology focuses on what is the same about two objects. It connects to geometry and algebra.

Knot theory is a branch of topology concerned with knots – closed loops in three-dimensional space. The unknotting number for a knot tells us how many cuts and reconnections are needed to unknot it. This is useful for studying DNA. Enzymes untie knots in DNA as part of the natural cell replication process, but sometimes it goes wrong and causes mutations. Biologists can understand the enzymes better by viewing flattened DNA strands under a microscope and using topological techniques to estimate the unknotting number.

WHY IT MATTERS
Topology lets us find connections between apparently unconnected areas, which allows us to use techniques from one area to solve problems in another

KEY THINKERS
Gottfried Leibniz
(1646–1716)
Leonhard Euler
(1707–1783)
Johann Listing
(1808–1882)

WHAT COMES NEXT
Category theory extends topological ideas and applies them to mathematical structures

SEE ALSO
Graph theory, p.76
Category theory, p.176

In
100
words

Topology studies the fundamental properties of objects without worrying about their exact shape, allowing us to find connections that would otherwise be hard to spot. Two objects are topologically the same if one can be smoothly deformed into the other without creating or closing any holes. For example, a sock made of infinitely stretchy material could be transformed into a blanket by squashing it flat and then stretching it out, so socks and blankets are topologically the same. You can't transform a T-shirt into a blanket without sewing up holes, so T-shirts are topologically different to both socks and blankets.

"Topology is the property of something that doesn't change when you bend it or stretch it, as long as you don't break anything."
Edward Witten, mathematical physicist

50

Möbius bands and Klein bottles

WHY IT MATTERS
Möbius bands and
Klein bottles have
interesting properties
that are studied in
topology

KEY THINKERS
August Möbius
(1790–1868)
Felix Klein (1849–1925)

WHAT COMES NEXT
The projective plane is
another one-sided
surface which can exist
without self-
intersecting only in
four dimensions

SEE ALSO
Topology, p.90

You can make a Möbius band by putting a twist in a strip of paper and joining the ends together. Without a twist, it would be a cylinder with two edges and two sides, but the twist makes the two edges of the original strip into one edge – you can follow all the way around it without leaving the paper.

You could make a Klein bottle by twisting a cylinder half inside-out before joining the ends together – but in three-dimensional space this is impossible unless the surface passes through itself. A Klein bottle can exist without passing through itself only in four-dimensional space. As a Möbius band has an edge (or boundary) it is an open surface, whereas a Klein bottle has no boundary and is a closed surface.

Klein bottle

A Möbius band is a 3D shape made from a strip joined with a twist, and a Klein bottle is a higher-dimensional version of the same idea. A piece of paper has a front and back; a sphere has an inside and outside. In both cases, the two sides are completely separate from one another. A Möbius band is like a one-sided piece of paper, and a Klein bottle is like a one-sided sphere. In either case, an insect walking on the surface could reach every part of the shape's surface without climbing over an edge or making a hole.

Möbius band

51

Trigonometry

Trigonometry is the study of the relationships between angles and lengths. The word literally means "measurement of triangles" (*trigon* is Greek for "three-angled"), but the concepts derived from trigonometry end up describing many aspects of mathematics and the world, even if they have nothing to do with triangles.

The basic functions involved in trigonometry are $sin(x)$ and $cos(x)$, or sine and cosine. These functions describe the coordinates of a point moving in a circle. For example, a point 30° around a unit circle, moving anti-clockwise from $(1,0)$, has coordinates $(cos(30°), sin(30°))$ which is approximately $(0.87, 0.5)$.

Some properties of the trigonometric functions change if you use different units for measuring the angles. Many important results are simpler and more useful if you use radians. Using Pythagoras' theorem (see page 98) on the-right angled triangle in the diagram gives us the first of many important facts (known as **identities**) concerning trigonometric functions: $cos^2(x) + sin^2(x) = 1$.

SEE ALSO

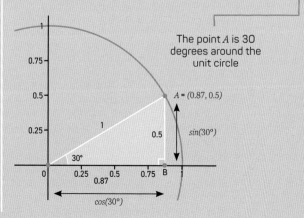

The point A is 30 degrees around the unit circle

$A = (0.87, 0.5)$

$sin(30°)$

$cos(30°)$

The trigonometric functions sine and cosine
describe a point moving in a circle. They can be
used to find measurements of sides and angles in
triangles (and therefore other shapes), and define a
third function $tan(x) = \frac{sin(x)}{cos(x)}$, short for *tangent*.
Plotting a graph of sine or cosine gives a wave
shape (a sine wave). These waves arise when
anything moves with **cyclic** or oscillating
behaviour – for example, a wheel, a bouncing
spring, or a sound wave. Trigonometric functions
are sometimes called the circular or elliptical
functions. They describe many real-world
situations and have deep connections
to many other parts
of mathematics.

The graph of a
sine wave

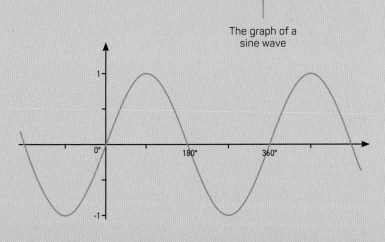

52

Radians

In everyday use, it's common to use degrees to measure angles. There are 360° in a full circle, which makes it easy to divide – 360 is a highly composite number (it has lots of factors).

The degree is an arbitrary unit defined to make measurements convenient. There is another unit for angles, the radian, which is based on a fundamental property of a circle. Radians make many ideas in calculus and other areas easier to express.

Trigonometric functions are intrinsically linked to circles, and they make more sense when used with radians. For example, the approximation $sin(x) \approx x$, used for small values, is only true when x is in radians. The **derivative** of $sin(x)$ is $cos(x)$ – but again, this is only true when x is in radians.

As calculus gets more advanced, working in degrees throws up multiples of $\frac{\pi}{180}$ all over the place; if you work in radians, these vanish and the calculations are much easier to work with.

A radian is a unit for measuring angles, usually given in terms of π. A full turn, which means turning all the way round to end up facing the same way as you started, is an angle of 2π radians, so $360° = 2\pi$ radians and 1 radian is approximately $57.296°$. One radian is the angle you move when travelling a distance the same length as the radius around the edge of a circle. This concept was first defined in the 1700s, but the name *radian* was not used until the 1870s. The radian is the SI unit for angle.

1

1

1 radian

1

An angle of 1 radian at the centre of a circle makes a distance the same length as the radius around the circumference

53

Pythagoras' theorem

Pythagoras' theorem relates the two short sides of a right-angled triangle to the longest side. Although it is named after Pythagoras, there is evidence that it was known long before his time, in Babylonian, Egyptian, Indian, and Chinese civilisations.

Although Pythagoras' theorem is stated in terms of squares of side lengths, it's not restricted to square shapes. If you draw a regular hexagon on each side of a right-angled triangle, where the side length of each hexagon is equal to that of the side it is drawn on, the area of the largest hexagon will be equal to the combined area of the other two. This works for any shape, as long as the three shapes are **mathematically similar**.

Hippocrates used this fact to show that the combined area of two lunes (crescent-shaped areas defined by two circular arcs) constructed on the short sides of a right-angled triangle is the same as the area of the triangle. Many people have tried to extend this method to construct a square with the same area as a given circle, but this was proven to be impossible by Ferdinand von Lindemann in 1882, when he proved that π is a **transcendental** number (see page 10).

The longest side of a right-angled triangle is called the hypotenuse. Pythagoras' theorem says that the square of the hypotenuse (a) is equal to the sum of the squares of the other two sides (b and c), or as an equation, $a^2 = b^2 + c^2$. The cosine rule extends Pythagoras' theorem to non-right-angled triangles: $a^2 = b^2 + c^2 - 2bc \times cos(A)$ (where A is the angle opposite the side a). A Pythagorean triple is a set of three **integers** that satisfy the equation of Pythagoras' theorem. For example, 3, 4, 5 is a Pythagorean triple because $5^2 = 3^2 + 4^2$. There are an infinite number of Pythagorean triples.

Pythagoras' theorem can be used to show that the combined area of the two shaded lunes is equal to the area of the triangle

b

c

a

54 Euclid's *Elements*

In the *Elements*, Euclid starts by defining five basic **axioms**, or **postulates**, and goes on to describe and prove many well-known constructions and theorems in geometry and number theory, including the golden ratio (see page 24), Pythagoras' theorem (see page 98), and the irrationality of the square root of 2 (see page 21).

The geometric constructions form a comprehensive guide to compass-and-straight-edge geometry, and have been reproduced in many forms, including famously by Oliver Byrne. In modern online editions they are given as animations showing each step of the construction, which are easier to follow than the original descriptions and bring the ideas to a new audience.

In **100** words

Euclid's *Elements* is a set of 13 books written by the ancient Greek mathematician, Euclid. Since its initial publication in around 300BCE, it has been translated into many languages and printed in thousands of editions. It was a key textbook in many schools until the early 20th century.

One reason it was so influential was its rigorous application of logic. It builds each theorem, construction, and idea step by step from those that have already been proved. Many of the ideas were not original to Euclid, but he was the first to collect them all together in one place.

55

Construction means creating an accurate distance, angle, or shape using geometric facts and properties. Classical construction uses a compass and a straight edge, starting with two given points a distance of 1 apart. This serves as a unit length to compare other distances to. Any new point used in the construction comes from the intersections of lines and circles. A number is constructible if it's possible to construct it as a distance between two points. All **rational numbers** and some **algebraic numbers** are constructible, but no **transcendental** numbers are, so it's not possible to construct a line with length π.

WHY IT MATTERS
Geometric construction was a forerunner of algebraic proof

KEY THINKERS
Ancient Greeks (1200–323BCE)

WHAT COMES NEXT
Euclid used compass-and-straight-edge constructions to prove the propositions in his *Elements*

SEE ALSO
Rational and real numbers, p.10
Euclid's *Elements*, p.100
Circle, p.104
Proof, p.167

Construction

The idea of constructing shapes and numbers using a compass and a straight edge was the basis of ancient Greek mathematics. Starting with two points, draw a circle centred on the first point, going through the second. Then draw another circle, centred on the second point, going through the first. The circles cross at two new points, each of which makes an **equilateral triangle** with the first two. This process also constructs a length of $\sqrt{3}$, an **irrational number**.

It's also possible to construct a regular pentagon. Doing so constructs the golden ratio, Φ (see page 24) – it's the distance between any two corners that aren't joined by a side of the pentagon. So Φ is an example of a constructible number.

There is no compass-and-straight-edge method to construct a regular heptagon (a seven-sided polygon), but you can construct one using origami. Origami construction uses the intersections of folds to create new points.

56

Conic sections

WHY IT MATTERS
These shapes turn up everywhere, but a good example is a satellite dish, which is a parabolic reflector (a three-dimensional shape with parabolic cross-sections) collecting signals and focusing them to a point

KEY THINKERS
Menaechmus (c. 380–320BCE)
Apollonius of Perga (c. 240–190BCE)
Johannes Kepler (1571–1630)

WHAT COMES NEXT
Studying conics and their properties has led to many other useful parts of mathematics – for example, the circular functions (see page 94) and the hyperbolic functions

Imagine a cone being sliced with a straight cut in some direction. What shape cross-sections could you create?

If you slice exactly horizontally this will create a circle, and if you slice slightly off the horizontal the result will be an **ellipse**. If you slice at exactly the same angle as the side of the cone, the outline of the shape will be a **parabola**, and if your slice is at a steeper angle you will make a **hyperbola**. A vertical slice through anywhere except the exact centre of the cone will make a particular type of hyperbola called a rectangular hyperbola.

These shapes are called the conic sections, or conics, and their shapes and equations are closely related to each other. Each one has important properties.

Astronomical objects (like planets and stars) travel around each other along conic sections, because of gravitational effects. In orbit, they travel along a circle or ellipse; otherwise they move along a hyperbola.

Trying to locate something using GPS or radio transceivers will result in a set of possible locations that lie along a conic section.

Conic sections are the five possible types of shape resulting from slicing a cone. We can class all conics as either **ellipses**, **parabolas**, or **hyperbolas**, since the circle is a special case of the ellipse, and the rectangular hyperbola is a special case of the hyperbola. As well as being cross-sections of a cone, they can be described geometrically and algebraically by the distance from a point (the focus) and a line (the directrix).

They arise in many abstract and real-life contexts, including falling objects modelled under gravity, which move along parabolas, and planetary orbits, which are ellipses and hyperbolas.

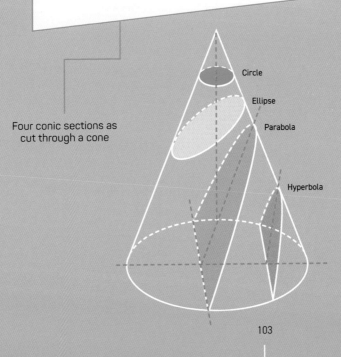

Four conic sections as cut through a cone

Circle

Ellipse

Parabola

Hyperbola

Circle

Circles are all around us, and are the most symmetrical possible flat shape. As a result they generate many useful properties: wheels that roll smoothly, gears that don't jam, and a metaphor for anything that repeats.

Circle theorems are mathematical results that relate the properties of angles and shapes in circles, and have many uses in geometry. Comparing the circumference of a circle to its diameter gives the mathematical constant π.

A point moving round a circle can be defined using trigonometric functions, showing the important links between circular geometry and trigonometry.

WHY IT MATTERS
Because of the importance of things that repeat (i.e. go in a "cycle", from the same root word), circles and their properties connect to almost every aspect of the real world and mathematics

KEY THINKERS
Ancient times, before recorded history
Euclid (c. 300BCE): Book 3 of his *Elements* was all about circles and their properties

WHAT COMES NEXT
In three dimensions, the collection of all points a fixed distance from a given point is a sphere. The circle and other closely related shapes arise from slicing a cone

SEE ALSO

In 100 words

The circle is the simplest two-dimensional shape, and one of the most important. It is the collection of all points at a fixed distance, the radius (r), from a given point, the centre. It can be represented algebraically as $x^2 + y^2 = r^2$. The distance around the outside is the circumference, with length $2\pi r$. A line joining two points on the circumference is called a chord – if it also passes through the centre, it is a diameter, with length $2r$. The area of a circle is πr^2.
The profound mathematical simplicity of the idea of a circle creeps into almost every mathematical topic.

In **100 words**

The two basic hyperbolic functions are $sinh(x)$ and $cosh(x)$. If we define a point with coordinates $(cosh(x), sinh(x))$ and vary the value of x, they will draw out a **hyperbola**. They can be defined in terms of exponential functions: $sinh(x) = \frac{1}{2}(e^x - e^{-x})$ and $cosh(x) = \frac{1}{2}(e^x + e^{-x})$. We can also define $tanh(x) = \frac{sinh(x)}{cosh(x)}$ the hyperbolic tangent (often pronounced "tansh"). The hyperbolic functions $sinh(x)$ and $cosh(x)$ behave in similar ways to $sin(x)$ and $cos(x)$ when differentiated or integrated and when forming **identities**: for example, $cosh^2(x) - sinh^2(x) = 1$. However, they also have important differences: they are not periodic (they do not repeat in a cycle) and their graphs look different.

Hyperbolics

A circle (or **ellipse**) can be described by using the circular functions $cos(x)$ and $sin(x)$ to define the coordinates of a point (see page 104 and page 94). In exactly the same way, a **hyperbola** (see page 102) can be described using two hyperbolic functions: $cosh(x)$ and $sinh(x)$, often pronounced as "cosh" and "shine".

Hyperbolic functions can be used to describe some everyday physical things, and they can help us solve differential equations. For example, the shape a hanging chain forms under gravity is a $cosh(x)$ curve, called a catenary. Rotating a catenary around an axis gives a surface called a catenoid, which is the surface joining two circles that requires the least material to build – you can see this shape in the large cooling towers of power plants.

WHY IT MATTERS
The shape of a hanging cable or chain is described by hyperbolic functions, for example the way a high voltage power cable will dip between pylons. Hyperbolic functions are also useful when solving calculus problems

KEY THINKERS
Vincenzo Riccati (1707–1775)
Johann Heinrich Lambert (1728–1777)

WHAT COMES NEXT
The hyperbolic functions can be related to their trigonometric equivalents using **complex numbers**

SEE ALSO
The complex plane, p.36
Conic sections, p. 56
Differential equations, p.64
Trigonometry, p.94

59

Distance metrics

When studying points in n-dimensional space, or more abstract notions like elements of a **set** of words, it's useful to have a notion of distance, called a metric. Defining a metric requires a few constraints: a metric must be consistent – it should give the same value no matter how it's measured (in the example shown, which uses the taxicab metric, all three routes involve six units of movement); the distance from a point to itself must be zero; and the distance via a third point must be the same or longer.

Our intuitive sense of distance corresponds to Euclidean distance, which is found using Pythagoras' theorem as the square root of the sum of the squares of the horizontal and vertical distances.

For non-physical spaces like collections of words, a metric called Hamming distance allows us to measure how different two points are: for example, HELP and HELD differ in only one place, but HELP and FISH differ much more. This idea is integral to error correction and data encoding.

A metric is a formalised definition of the distance between two points. The Euclidean metric corresponds to our usual notion of distance: the length of a straight line between two points. Other metrics include the taxicab metric, or Manhattan distance, in which we imagine driving a taxicab through a grid of streets; it is given by the sum of the horizontal and vertical distances. We can also consider distances between objects other than points in space: the Hamming distance defines the distance between two strings of letters of the same length as the number of positions in which they differ.

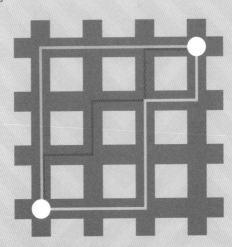

In the taxicab metric, these two points are six units apart no matter which route you take

60

Gaussian curvature

Gaussian curvature is a measure of how a 3D object curves. It is an intrinsic property of a surface and can't be changed – if you pull on the sides of a saddle shape, which naturally curves up and down, the front and back twist to compensate. It can also be useful in real life – if you bend a slice of pizza, its zero curvature property means it won't droop!

Shapes with different curvatures possess different geometrical properties. In **spherical geometry**, a triangle's angles can add up to more than 180 degrees, and in **hyperbolic geometry**, straight lines cannot be parallel as they must always meet somewhere.

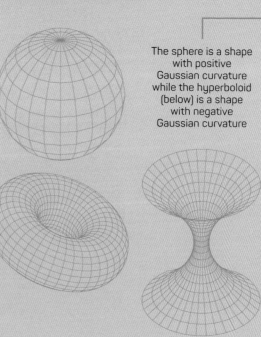

The sphere is a shape with positive Gaussian curvature while the hyperboloid (below) is a shape with negative Gaussian curvature

Gaussian curvature is a measure of the way an object curves in two **perpendicular** directions (at right angles). A sphere has positive Gaussian curvature: from a point on the top, it curves downwards in all directions, while a hyperbolic shape like a saddle has points with negative curvature: it curves down at the sides, but up at the ends.

A flat piece of paper has zero Gaussian curvature, meaning if it curves one way it must stay flat in the other; roll paper into a tube, and it becomes difficult to bend in the other direction without tearing or crumpling.

61

Vectors

Vectors can be thought of as arrows – starting from a point, extending in a particular direction with a given length. They are commonly represented as a list of numbers, representing the distance moved in each dimension the vector is defined in. For example, $(1, -2)$ is a two-dimensional vector representing moving across one unit and down two units.

Vectors can be used to model physical forces and movement – representing motion in space, or the **velocity** of a moving object. Vector fields – consisting of a vector attached at each point in a space – can model air currents, flows in substances, or even hair.

In
100
words

A vector is a geometric object possessing length and direction. Vectors are commonly considered in two- and three-dimensional space, and can be defined by giving the coordinates of the endpoint of the vector from the **origin**. Scaling (multiplying a vector by a number) will change its length but not direction.
Vectors can be combined by addition – positioning the start of one vector at the end of another, and considering the vector from the start point of the first to the endpoint of the second. Vector spaces are algebraic structures, consisting of vectors under the operations of scaling and vector addition.

Applied Maths

Mathematics is concerned with ideas – about patterns, connections, numbers, and representations – but nothing in this description requires these ideas to be useful. Many mathematicians have celebrated the abstract character of maths as part of its beauty, setting it apart from other, more applied subjects. The fact remains, however, that many ideas in mathematics are also extremely useful, and have been used by humans to build and change the world around them for thousands of years.

While pure mathematics refers to mathematical ideas studied independently of any usefulness outside of maths, applied mathematics is a label for all maths connected with describing, predicting, or controlling our world in some way – whether physical effects, statistical outcomes, or computational techniques.

Distinctions between Applied Maths and Pure Maths are blurry, and the labels overlap: mathematical ideas developed as purely abstract concepts find applications in the real world, and ideas arising from applied mathematics might be pursued for their own sake, independent of any industrial application.

62

Modelling

Mathematicians often try to represent aspects of the real world using mathematics so that they can understand how they work and make predictions. This is the process of modelling, the heart of all applied mathematics.

The word *model* originally comes from a French word for a set of scale drawings for a building. In that sense – an attempt to represent something real in a different or simplified way – it describes how we use the word for both a fashion model (a representation of how clothes might look on a person) and a mathematical model (a representation of how a real-world process might work). These representations are useful ways of thinking about the real thing, rather than exact copies.

For example, modelling how a thrown object behaves might start by assuming that there is no air resistance and gravity acts constantly downwards. For something like a stone, this is a good model, because air resistance has little effect. However, to model a falling parachutist, we would need to include air resistance to make a useful prediction.

Humans have been representing the real world by simplified abstractions since before recorded history – a cave painting could be the first example of a model.

WHY IT MATTERS
Getting the balance right between a model that is simple enough to solve but detailed enough to represent the real world is hard. Too simple and it does not give trustworthy predictions; too complicated and it becomes impossible to get any predictions at all

KEY THINKERS
George Box (1919–2013) was a British statistical modeller and mathematician

WHAT COMES NEXT
In order to turn our intuition about the real world into mathematics, we often use algebra and equations (see pages 44–46) to represent our ideas, and calculus (see page 58) to describe how things change

SEE ALSO
Differential equations p.64
Mechanics, p.114
Monte Carlo simulation, p.116
Discrete maths, p.118
Statistics, p.142

A mathematical model is a representation of a real-world process with a mathematical idea. Mathematical models can help us solve problems or predict outcomes. Typically, **assumptions** are made to simplify the problem and allow a solution to be found, then the model can be upgraded to include more detail and reflect the real process more accurately. This process of repeatedly proposing, solving, checking, and upgrading the model is known as the **modelling cycle**. Many mathematical models of the real world use differential equations to set up the model. These are then solved and the outcomes compared with the real data.

"All models are wrong, but some are useful."

George Box, statistician

63

Mechanics

WHY IT MATTERS
Understanding and
predicting the physical
world is one of our
most used applications
of mathematical tools,
whether we want to
predict the weather,
design a safe car, or
fly to the moon

KEY THINKERS
Galileo Galilei
(1564–1642)
Isaac Newton
(1643–1726)
Max Planck
(1858–1947)
Albert Einstein
(1879–1955)

WHAT COMES NEXT
Joining relativity and
quantum mechanics
together in a unified
theory is a major
goal in applied
mathematics and
theoretical physics
which is often referred
to as the "theory of
everything"

SEE ALSO
Modelling, p.112

Mechanics is the area of mathematics used to model moving objects, forces, and masses. It is the part of maths that most overlaps with the physical sciences, or physics.

Our models of everyday objects are based on Isaac Newton's three laws of motion, paraphrased as follows:

- An object remains at rest, or at a constant speed, unless acted upon by a force.

- The force acting on an object is equal to the acceleration multiplied by the mass.

- If two objects exert forces on each other, the forces have the same size but opposite directions.

It is important to remember that our attempts to represent the real world are necessarily only models and Newton's laws do not always hold true, but they have nevertheless proved remarkably effective at predicting our physical world on everyday scales.

> **"Mechanics is the paradise of the mathematical sciences, because by means of it one comes to the fruits of mathematics."**
> **Leonardo da Vinci**, polymath

In 100 words

Mechanics is the mathematics of forces, matter, and movement. It gives us tools to predict how things accelerate, fall, slip, slide, and turn.

Classical mechanics – often called Newtonian mechanics after Isaac Newton – models the motions and forces of everyday physical objects and provides useful and accurate predictions when dealing with processes at a reasonable size or speed. For massive objects and high speeds (near the speed of light), Einstein's theories of **special and general relativity** become more accurate, and at very small scales (atomic level and below), we need the theory of **quantum mechanics** in order to make accurate predictions.

64

Monte Carlo simulation

Sometimes we can't actually solve an equation we'd like to, or calculate an expected outcome for a probability question.

However, we can run a simulation: setting up some initial rules (the model) on a computer and letting it run repeatedly to see what happens.

For example, rolling and adding the values of two 6-sided dice is a familiar but fundamentally unpredictable experiment. In this case, it is relatively easy to calculate the expected distribution of outcomes. But we can also estimate the probability of each outcome by simply repeating the experiment. The blue bars on the images show the proportion of outcomes from experimental tests after 10 rolls, 100 rolls, and 1,000 rolls. Even after only 1,000 tests, the blue bars provide a decent approximation of the true expected outcomes.

Stanislaw Ulam and John von Neumann pioneered these sorts of probability simulations during the Manhattan Project in the 1940s (developing nuclear weapons) as computing technology developed enough to make it possible. Their work was a military secret and it needed a code name: "Monte Carlo" (after the famous Monte Carlo Casino in Monaco) was chosen – an appropriate name for a method heavily relying on randomness and chance.

When we can't solve a problem analytically we can repeatedly run experiments to eventually provide enough data for us to be fairly confident about what outcomes to expect.

This is called a Monte Carlo simulation. It is a surprisingly simple and effective concept that can solve otherwise intractable problems, but it requires a lot of data, so these simulations have only become feasible with computer technology.

Monte Carlo simulations can be quickly implemented on computers using a source of randomness. Computers can generate **pseudo-random numbers** for this using known chaotic systems, which appear to behave randomly.

1: expected distribution, 2: proportions in blue after 10 rolls, 3: after 100 rolls, 4: after 1000 rolls

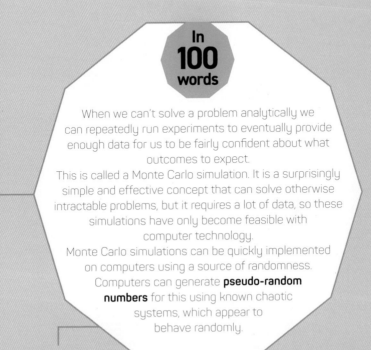

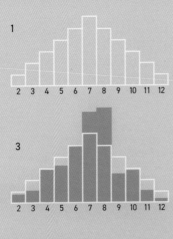

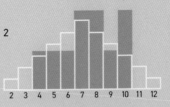

65

Discrete maths

Discrete mathematics is a label for mathematical ideas with distinct steps or parts – discrete chunks (like counting sheep) – as opposed to continuous measurements (like measuring height). These ideas include graph theory, algorithm design, and problems involving counting and **integers** (see page 9).

Discrete maths has gained importance with the rise of computing, which operates in a discrete way (discrete steps of an algorithm, storing information as discrete 1s and 0s). As a result, it often deals with problems in computer science, such as complexity, computability, cryptography, and coding.

WHY IT MATTERS
Ideas of efficiency, complexity, and security are vital in the modern world, and these ideas are effectively handled by the concepts in discrete mathematics

KEY THINKERS
Leonhard Euler
(1707–1783)
Alan Turing
(1912–1954)

WHAT COMES NEXT
Modern ideas of *what* we can do with computers, how we can do it, and how efficiently we can do it are all consequences of discrete mathematics

SEE ALSO

In
100
words

Discrete maths is the study of any mathematics represented by discrete or distinct parts, rather than continuous measures. This distinction has different vocabulary depending on the area being studied, such as "digital vs analogue" (often in electronics and computing) or "countable vs uncountable" (in **set** theory and **number theory**). Graph theory, discrete optimisation, and the efficiency and design of algorithms are just a few examples of areas that involve discrete mathematics. The use of computer algorithms that implement cryptography, which is essential to keep information secure, is one of the most important applications of discrete mathematics in the modern world.

From simple ciphers like alphabetical shifts or number codes, techniques and devices to implement secret codes have advanced alongside mathematical ideas. Modern cryptography uses powerful **number theory** techniques to create encryption secure enough to repel attacks by computers. Often this is based on a one-way function: one that takes a lot more computing power to undo than to do, like finding the factors of the product of two large primes. Meanwhile, mathematical key-exchange protocols, which involve creating and sharing secret keys between two parties, can be used to transmit encryption keys on an insecure channel with no chance of interception.

Cryptography

Since ancient times, people have used secret codes to communicate, particularly during times of conflict or when secrecy was useful. Cryptography involves encoding or otherwise obscuring a message so that it can only be understood by its intended recipient.

Mathematical techniques from **number theory** have allowed us to create increasingly secure encryption, and it's now used as the basis for bank transactions, digital signatures, and secure data storage.

No system is 100% secure – to decrypt the message, you need to receive a **decryption key.** This creates vulnerabilities, but modern key exchange protocols can be used to send keys securely.

WHY IT MATTERS
Cryptography underpins all of modern communications, finance, and security

KEY THINKERS
Julius Caesar (100–44BCE)
RSA encryption: Clifford Cocks (1973) and Ron Rivest, Adi Shamir, Leonard Adleman (1976)
Key exchange: Ralph Merkle, Whitfield Diffie, Martin Hellman (1976)

WHAT COMES NEXT
Quantum computing may mean that processing speeds become so fast, classical cryptography methods become useless and **quantum cryptography** becomes the new standard

SEE ALSO
Prime numbers, p.16
Modular arithmetic, p.33

67

Algorithms

WHY IT MATTERS
Algorithms provide
defined procedures
that can be followed
by anyone or by a
computer

KEY THINKERS
Muhammad ibn Musa
Al-Khwarizmi
(9th century)
John Napier
(1550–1617)
Ada Lovelace
(1815–1852)

WHAT COMES NEXT
Algorithms for simple
tasks can be used as
steps in more complex
tasks, leading to the
ability to solve much
more complicated
problems

SEE ALSO
Basic operations, p.14
Logarithms and *e*, p.22
Linear
programming, p.132

Any time you follow a defined sequence of steps to do a particular task, you are performing an algorithm – this might be the sequence of calculations for long multiplication, the steps in a recipe to make a lasagne, or even the process to train a dog. Some of these algorithms are better defined than others, and the results depend not only on how well the algorithm is followed but also on the inputs – a training algorithm that works for a sensible collie might have limited use on an excitable spaniel. Choosing the best algorithm for a task is essential.

All computer programs are algorithms. This important fact was recognised by Ada Lovelace, referring to the potential of Charles Babbage's analytical engine – a design for a steam-powered calculating machine. Lovelace's notes on the machine outlined how it could be used to manipulate symbols and letters, as well as numbers, and included a diagram of an algorithm for calculating the **Bernoulli numbers**, commonly cited as the first computer program.

Even modern AI chat programs are based on algorithms, but because they are so complex, with built-in randomness and huge and varied inputs, they give the illusion of natural thought.

An algorithm is a set of instructions, performed on given inputs, resulting in specific outputs. Algorithms are related to functions: functions define *what* happens; algorithms define *how* it happens. For example, the function "divide by" takes two numbers as inputs and gives one number, the quotient, as the output. To achieve this, you could count how many times you can subtract one number from the other, or use the long division method, or follow the division procedure on Napier's bones (a set of rods with numbers engraved on them). These are all examples of algorithms that implement the division function.

"The Analytical Engine weaves algebraic patterns, just as the Jacquard loom weaves flowers and leaves."
Ada Lovelace, mathematician and early computer scientist

68

Complexity and sorting

Algorithms can be broken down into steps, but an important question we might ask is *how many* steps? Some algorithms are extremely efficient and can be run in just a few nanoseconds on a powerful computer, while others take more computation power to crunch through. It's also important to consider how much memory storage a process needs. Keeping both computation time and memory storage small is a goal for programmers and algorithm designers.

Since some algorithms might take longer as the size of the input grows (for example, the length of the list to be sorted, or the number of digits in a number), it's important to have a measure of how this changes. In Big O notation, $O(n)$ is used to indicate a function like addition, which takes n steps to run for n-digit number inputs, while multiplication by hand is $O(n^2)$, meaning roughly n^2 steps are needed.

Types of sorting algorithms include bubble sort, **quicksort**, **insertion sort**, and **merge sort**. The most efficient sorting algorithms, including the merge sort, are $O(n \times log\ n)$ (see page 22). Computing digits of π by hand is $O(n^2 \times log\ n)$, and matrix multiplication is $O(n^3)$.

In 100 words

Computational complexity measures how long a process takes to run. Complexity is often stated in terms of n, the input size, using Big O notation and a function of n.

Sorting algorithms can be used to sort lists of inputs. Simple examples include a bubble sort, which compares each pair of elements in the list and puts the larger one to the right, repeatedly running through the whole list until it is sorted. While this will always return a correct ordering, it is inefficient and can take as many as n^2 steps to sort n items – its complexity is $O(n^2)$.

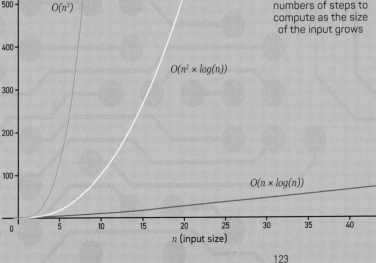

number of steps

$O(n^3)$

$O(n^2 \times log(n))$

$O(n \times log(n))$

Different algorithms take different numbers of steps to compute as the size of the input grows

n (input size)

69

P vs NP

P (polynomial time) and NP (nondeterministic polynomial time) are classes of mathematical problems, based on the numbers of steps involved in computing their solutions.

Problems in class P have a quick way to find a solution, and problems in class NP have a quick way to check a solution. For example, finding a **subset** of numbers from a given **set** that add to a particular total could take many calculations, but only one calculation is needed to check whether a given set of numbers adds to that total – this is like an NP problem. All problems in P are also in NP. It's possible that there are quick ways of finding solutions to NP problems, as well as checking them, but that no one has yet found any.

Given a map of cities and the distances between them, the travelling sales rep problem involves working out whether there is a route, shorter than a given length, that visits each city exactly once. It's simple to check that a proposed solution is correct (so it's in NP), but there's no known quick way of finding a solution (so we don't know if it's in P).

In
100
words

P vs NP is the open question of whether every NP problem is also an P problem. P stands for polynomial time: the number of steps to *find* a solution is a polynomial function (see page 48) of the number of inputs (n).

NP stands for nondeterministic polynomial time. The number of steps to *check* a solution is a polynomial function of n (e.g. n^2), but the number of steps to *find* a solution might be, for example, an exponential function of n (e.g. e^n), which increases much faster. If P = NP, then problems in NP are also in P, and can be solved in polynomial time.

70

Graph theory

WHY IT MATTERS
Graph theory is used to model many real-world situations, including transport networks, the spread of rumours in sociology, and nervous systems in biology

KEY THINKERS
Leonhard Euler (1707–1783)

WHAT COMES NEXT
Tools from graph theory are used to study other structures in, for example, group theory (see page 76) and **knot theory**

SEE ALSO
Discrete maths, p.118
Four colour theorem, p.128
Traversability, p.130
Linear programming, p.132

Graph theory, the study of **sets** of points and how they are connected, can simplify situations where the connections between objects are important. The London Underground map is an example of this type of graph: although the information about the geographical location of the stations isn't accurate, it does tell you exactly how each station is connected to the others. Removing unnecessary information makes it easier to work out routes from one station to another. If it included an estimate of travel time for each connection, it would be a **weighted graph**, from which you could work out which route is faster.

The name *graph* is related to the word *graphic* and captures the idea of representing important information with visual information rather than just data.

We can draw graphs in different ways depending on what we want to know about them. The first two diagrams look different, but they represent the same graph – in the second diagram it has been rearranged so that no edges cross over each other, but the vertices are connected in the same way. The third diagram has an extra edge, so it represents a different graph – it can't be rearranged to have no crossings.

In the context of graph theory, a graph is made up of a set of points (called vertices) and a set of edges; each edge connects two points. A planar graph can be drawn on a flat surface without any edges crossing each other, and a graph in which every vertex is connected to every other vertex is called a complete graph. Extra information such as direction or cost can be added to the edges of a graph; this allows them to be used in **optimisation problems** such as finding the fastest route between locations (see page 132).

Edges

Vertices

71

Four colour theorem

In 1852, Francis Guthrie noticed that he needed at least four colours to colour the counties of England without any neighbouring counties being the same colour. He made a conjecture that no map would ever need more than four colours to produce the same effect; this became known as the four colour conjecture. It took 124 years before it was finally proved, with the help of computers, in 1976. The **proof** was controversial; although it is now accepted, many mathematicians denied it at first because it could not be checked without a computer.

> "Until an easily understood proof can be found – a proof without the aid of computers – the Four Color Theorem remains a heuristic that represents a 'Holy Grail' for many mathematicians."
>
> **Teena Andersen**, maths educator

The four colour theorem says that
you need at most four colours to colour
the regions of a map so that no two neighbouring
regions are the same colour. We can model this idea
mathematically using a graph: each region is represented
by a vertex, and each border between two regions by
an edge. The number of colours needed to colour every
vertex so that no two connected vertices are the
same colour is called the chromatic number.
The maximum of four applies only to
planar graphs – graphs that can
be drawn on a flat surface
without any edges
crossing.

**"Its computer aided proof has
forced mathematicians to
question the notions of proofs
and mathematical truth."**
Andreea S. Calude, data linguist

72

Traversability

WHY IT MATTERS
Eulerian paths and
circuits have many
applications, including
in bioinformatics and
circuit design

KEY THINKERS
Leonhard Euler
(1707–1783)

WHAT COMES NEXT
A related, but different,
idea in graph theory is
the Hamiltonian path,
which visits each
vertex exactly once

SEE ALSO
Graph theory, p.126
Four colour
theorem, p.128
Linear
programming, p.132

The first known mention of graph theory is in a paper by Leonhard Euler, who developed it to analyse a problem about a walk around the town of Königsberg (now called Kaliningrad). Four regions of Königsberg were separated by rivers; in Euler's time, the regions were connected by seven bridges. The story goes that townsfolk on a walk around town would try to cross each bridge exactly once, but no one had ever found a way to do it.

Euler settled the question by modelling the town as a graph, with the land masses represented by vertices and the bridges represented by edges (see page 126). This more abstract diagram showed him that any path using all the edges once would have to go through each vertex, except the first and last, an even number of times – because each time you travelled towards a vertex, you would then have to also travel away from it. But every land mass in Königsberg had an odd number of bridges connected to it, so crossing every bridge exactly once would be impossible.

A graph is traversable if it's possible to find a route along the edges that goes over every edge exactly once. This type of route is called an Eulerian path; if it also starts and ends at the same vertex, it's called an Eulerian circuit. You can tell whether a graph is traversable by counting the number of edges connected to each vertex. If they all have an even number of edges, it's possible to construct an Eulerian circuit. If exactly two of them have an odd number of edges, then it's possible to construct an Eulerian path.

A diagram of Königsberg's rivers with the land masses represented by vertices and the bridges represented by edges

Linear programming

Linear programming is often used to solve network flow problems modelled by graphs with a cost assigned to each edge (called **weighted graphs**). One such system revolutionised water planning in the UK after severe droughts in 1995. Software translates data on rainfall and treatment-work capacity into a linear programming problem, then solves it to find the most efficient way to meet demand. Companies can run the model on historical data to show them where to increase capacity to cope with worst-case scenarios.

WHY IT MATTERS
Linear programming is used to help decision-making in many industries, including the water industry, computer networking, and banking

KEY THINKERS
Tjalling Koopmans (1910–1985)
Leonid Kantorovich (1912–1986)
George Dantzig (1914–2005)

WHAT COMES NEXT
If the objective function isn't linear – for example, if there are economies of scale so that the profit per item increases with the number of items – then non-linear programming techniques must be used

SEE ALSO
Equations and inequalities, p.46
Algorithms, p.120
Graph theory, p.126

In **100** words

Linear programming is a method for finding the best outcome, subject to given constraints. For example, a baker might want to know which products will maximise profits given the ingredients available. We model this by creating an **objective function** for profit, where variables represent the number of each product type. Inequalities represent the constraints – the total amount of each ingredient used must be less than the amount available. We then use an algorithm to find the values of the variable that maximise the objective function while satisfying all the inequalities. Simplex is an efficient algorithm for solving linear programming problems.

A dynamical system arises when a function which gives outputs of the same type as its inputs is **iterated** – the output is put back in as the input, repeatedly, to observe long-term behaviour. A fixed point is a value which, when used as the input to the function, gives the same output. Making a small change to the input value at a fixed point lets us determine if it is unstable, like a ball balanced on a hill, which will roll away, or stable, like a ball at the bottom of a hole, which will return to the fixed point.

Dynamical systems

Dynamical systems allow us to study systems which change over time – from swinging pendulums to weather patterns. This kind of system can model many real-world situations, e.g. population dynamics, where changes to the size of a population depend on many factors – including the current population size. Dynamical systems can also model fluids – liquids and gases – in which each particle obeys rules about how to move relative to the ones around it, with the rules defined by a function.

Dynamics is often studied using differential equations, since these can be used to describe rates of change.

WHY IT MATTERS
Dynamical systems have applications across science, engineering, economics and medicine. Even the rise and fall of historical empires can be studied using these ideas

KEY THINKERS
Henri Poincaré
(1854–1912)
Oleksandr Mykolayovych Sharkovsky
(1936–2022)
Ali Hasan Nayfeh
(1933–2017)

WHAT COMES NEXT
Some dynamical systems exhibit a type of behaviour called chaos, which makes it harder to predict long-term behaviour

SEE ALSO

The logistic map

WHY IT MATTERS
The apparently random nature of the outputs of the logistic map in the chaotic region mean it can be used to generate pseudo-random numbers (see page 116), as can other chaotic functions

KEY THINKERS
Leonhard Euler Pierre François Verhulst (1804–1839)
Robert May (1936–2020)
Mitchell J. Feigenbaum (1944–2019)

WHAT COMES NEXT
Mitchell Feigenbaum noticed a persistent pattern in the ratios between the intervals where the logistic map doubles its cycles. He then showed that this ratio is the same for any similar chaotic function, and it is now named after him: Feigenbaum's number ≈ 4.6692

The logistic map is a simple population model, used to predict how populations will grow or shrink based on a reproduction rate, r.

Different outcomes are observed for different values of r, and all of these can be seen in real population data: for $0 < r < 1$, everything dies out; for $1 < r < 3$, the population stabilises; and for $3 < r < 3.4$, the population oscillates between two values.

For larger values of r, up to 4, the outcomes become surprisingly complicated and unpredictable. The diagram of these outcomes is often called a bifurcation diagram.

In 1970, biologist Robert May pointed out the surprising behaviour of this simple equation for large values of r. This was the first known appearance of what we now call chaos (see page 136).

Further study of this chaotic region led to a new area of mathematics called **nonlinear dynamics**, which is now used to model much of the real world, as well as producing beautiful fractals.

Mathematicians have shown that this sort of chaotic behaviour is entirely normal, rather than a surprising exception. We use it when modelling all sorts of real-world phenomena, including in weather forecasting.

The logistic map is a function which can be **iterated** to model simple populations: $f(x) = r\,x(1-x)$. Here, x represents a population size between 0 and 1, and r is the reproduction rate. The result is substituted back into the function to find the next generation's population size. Changing the starting population and the growth rate lets us model different outcomes for populations over time.

Study of the logistic map for larger values of r led to the first observations of chaotic behaviour, which in turn led to the creation of new branches of mathematics called chaos theory and dynamical systems.

The long-term outcomes of the logistic map model for different values of r

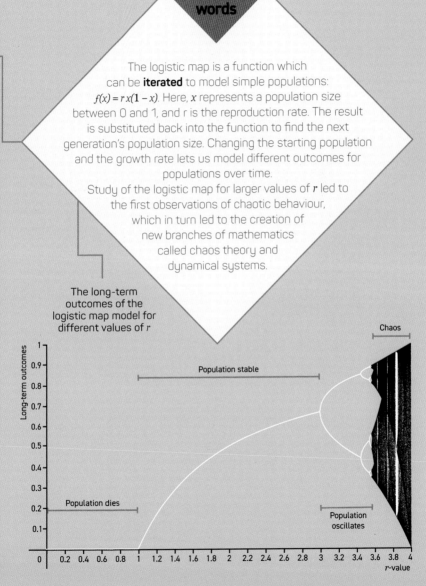

76

Chaos

While many think of the word "chaos" as implying uncontrollable, unpredictable behaviour, the mathematical definition has a subtle difference. Chaos is a property of dynamical systems, in which functions are applied repeatedly.

The evolution of these systems over time is by definition extremely predictable: if I tell you the initial input, you can calculate the result of applying the function a hundred times and give a precise answer.

Chaos occurs when we think about making a small change to the initial input. For some systems, this will result in a small, predictable change to the future state, but in chaotic behaviour it might change drastically.

Chaos theory concerns deterministic systems with sensitivity to initial conditions: a small change in the input might result in a large, unpredictable change to the output.

It's sometimes called the butterfly effect: a butterfly flapping its wings might cause changes in air pressure and weather systems resulting in tornadoes elsewhere. Within a dynamical system, we observe periodic orbits – **sets** of points which, if one is used as an input, the system will visit each of the others in a loop and return to the first. Chaotic systems contain periodic orbits of every period, filling the whole space of possible inputs.

"**Lorenz saw more than randomness embedded in his weather model. He saw a fine geometrical structure, order masquerading as randomness.**"
James Gleick, author

Fractals

Fractals are intricate mathematical diagrams or objects which have infinite levels of detail, often with a pattern repeating itself at different scales.

Some fractals can be constructed directly – the Sierpiński triangle is made by repeatedly removing the central section from an equilateral triangle, then repeating this process on each of the remaining equilateral triangles, and so on.

Others arise as visualisations of the properties of dynamical systems. The Mandelbrot **set** describes the behaviour of the complex function $f(z) = z^2 + c$, for each possible point c in the complex plane, using colour to indicate how many iterations of that function are needed to escape to infinity.

A fractal is usually a shape or image with some self-similarity – containing smaller versions of the whole pattern within it. Fractals tend to have a complex, detailed structure at every possible scale level, and are generated by iterating a process infinitely. They can be used to visualise the long-term behaviour of dynamical systems.

Fractals have interesting mathematical properties, and can be considered as not necessarily one-, two-, or three-dimensional, but somewhere in between. For example, the Sierpiński triangle is approximately 1.58-dimensional, and has an infinite length perimeter but zero area. The name "fractal" comes from this idea of fractional dimension.

Stages in the iterative process to create a Sierpiński Triangle

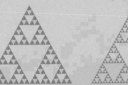

Probability and Statistics

Applied mathematics takes many different forms, but we almost always encounter the problem of determining how well the mathematical models compare with the real world. To find out about this comparison we have to collect data from the real world, and then process and analyse that data. Statistics, as the label for this process, deserves a chapter to itself.

Originating in the study of populations, statistics allows us to study small samples and make inferences from them about the whole group. These techniques and tools can also use raw data to draw conclusions about the past or present, and make predictions about the future, and are becoming increasingly important in the modern world, where vast amounts of data are generated every day about every conceivable thing.

Probability, the mathematics of uncertainty, is vital in our attempts to predict the world. It developed, in part, from discussions on how to make a game objectively fair for all the participants, and is a useful reminder of how even apparently frivolous concepts develop into world-changing ideas. This chapter includes descriptions of some particularly interesting games, and the area of mathematics now called game theory.

In 100 words

Probability measures the likelihood of an event occurring. It is expressed as a value between 0 and 1, which can often be written as a fraction or percentage. An event with a probability of 1 is certain to happen, and a probability of 0 means it will definitely not happen.

We can calculate the probability of two events with known probabilities both happening – if they are **independent** (neither event influences the other) we multiply the probabilities together. Two coin flips, each with probability ½ of returning heads, have a ½ × ½ = ¼ probability of both coming up heads.

Probability

We often encounter probability in everyday life – the chance of rain in the weather forecast, a dice roll in a board game, gambling odds, or a medical prognosis. Some probabilities can be calculated precisely – like the chance of picking a pair of matching socks given the number of each colour in the drawer – whereas others can only be approximated based on past evidence or a highly complicated model of interacting factors.

In science and statistics, probability is used to measure the meaningfulness of experimental outcomes, and actuarial scientists use complicated probability forecasts to determine insurance rates.

WHY IT MATTERS
Probabilistic methods are used throughout science. For example, they are used for modelling the spread of diseases and for economic, financial and weather systems

KEY THINKERS
Gerolamo Cardano (1501–1576)
Christiaan Huygens (1629–1695)
Abraham de Moivre (1667–1754)

WHAT COMES NEXT
When two events are not independent, we can use Bayes' theorem to calculate the combined probability

SEE ALSO

79

Statistics

Statistics is an overall term for anything to do with gathering, representing, and interpreting **data**. It covers a huge range of ideas – from designing sensible questionnaires and presenting evidence for new claims (hypothesis testing) to far-reaching theorems about the way all data samples are likely to be distributed (the **central limit theorem**).

The word originally comes from a German word for data about a country or "state", and this is a reminder that data are always *about something* – like a country – and rarely just abstract numbers.

Statistical methods eventually developed from vague descriptions into a mathematical science, particularly during the late 19th and 20th centuries. A key example in graphical communication of data was Florence Nightingale's reports of health issues during the Crimean War (see page 144). It is important to recognise that many of the ground-breaking ideas emerged from scientists like Francis Galton, Karl Pearson, and Ronald Fisher, who were studying how inheritance and genetics affect population characteristics. They used some of their new and effective statistical tools to promote racist ideas like eugenics, which have since been discredited. It is a sobering reminder that statistical tools are separate from the messages they are used to promote, and that data can be both used and abused.

WHY IT MATTERS
The modern world generates and stores vast amounts of data every day. With computer technology and mathematical statistics, these large data sets can be analysed to give valuable information

KEY THINKERS
Al-Kindi (801–873CE)
Florence Nightingale (1820–1910)
Francis Galton (1822–1911)
Karl Pearson (1857–1936)
Ronald Fisher (1890–1962)

WHAT COMES NEXT
The central limit theorem is a remarkable statement about how averages of random samples from data will almost always form a normal distribution, whatever the original distribution looked like

"Facts are stubborn things, but statistics are pliable."
Mark Twain, writer

In
100
words

Statistics is the discipline of gathering, representing, and interpreting **data**.
The concepts covered by the term "statistics" are wide ranging and include:

- surveys, censuses, and sampling
- graphical representation of data
- averages and measures of **spread**
- correlation between sets of data
- hypothesis testing and significance
- probability distributions and **random variables**.

Statistics is intertwined with probability, because our expectations of how data is distributed overlaps with our handling of the data. The concept of significance is important for how data from research is combined with probability theory to justify whether a result is important or valid.

80

Statistical diagrams

Statistical **data** can be complicated and difficult to interpret, so statistical diagrams are used to display the key points clearly.

Florence Nightingale was a pioneer of statistical diagrams. During the Crimean War, she collected statistics on the causes of death in soldiers in the military hospitals where she worked. She developed a type of diagram called a polar area diagram, where the size of different coloured wedges represented the number of deaths by different causes. The diagrams made it immediately clear that preventable disease from poor hygiene was by far the most common cause of death. Nightingale collaborated with journalist Harriet Martineau to publish her findings; as well as being instrumental in reforming military sanitary policies, her approach to statistical diagrams transformed the way in which statisticians communicate their findings.

Unfortunately, statistical diagrams can be used to mislead. In bar charts, the scales of the axes can emphasise or diminish differences – on an axis that goes from 90% to 100%, the difference between 94% and 97% looks huge, but if the axis goes from 0% to 100% then the bars will be close to the same height. Misleading scales are often used to emphasise a particular political point.

Statistical diagrams convey statistics quickly and effectively. They are also used as analytical tools – viewing **data** graphically can make it easier to see trends and proportions.

There are many different types of statistical diagrams. A bar chart uses the heights of rectangular bars to show different quantities; a histogram is similar, but with quantities represented by the areas of the bars. A pie chart uses the sizes of slices of a circle to show the proportions of a data set in different categories. Scatter graphs use points on coordinate axes to show how different values behave relative to each other.

An example of one of Nightingale's polar area diagrams

April 1855 to March 1856

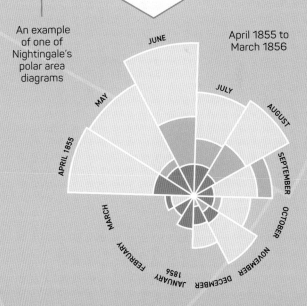

JUNE

JULY

MAY

AUGUST

APRIL 1855

SEPTEMBER

OCTOBER

MARCH

NOVEMBER

FEBRUARY

DECEMBER

JANUARY 1856

81

Bayes' theorem

While some events in probability can be considered independent as their outcomes do not influence each other, most probabilistic modelling needs to take into account dependent probabilities – when the outcome of one event affects the chance of another.

Bayes' theorem allows us to make calculations about such probabilities. For example, medical tests are rarely 100% accurate, so if someone tests positive for a disease, we can use Bayes' theorem to take into account the accuracy of the test and calculate the probability they actually have the disease. We could denote the event "having the disease" as A and "testing positive for the disease" as B. The probability of B depends on the status of A.

Imagine a disease affects 5% of the population, but the test has a 1% chance of giving a **false positive** or a **false negative**. Bayes' theorem tells us that if a representative subject does test positive, they still have around a 19% chance of not having the disease.

Since the tests are not 100% accurate, Bayes' theorem gives us a way of calculating how to update our beliefs when we get new data.

For two events A and B, we can calculate the probability of A happening given that B has happened, denoted $P(A|B)$, as: $P(A|B) = \dfrac{P(A \cap B)}{P(B)}$

Here, $P(A \cap B)$ is the probability that both A and B happen, and it is divided by the probability of B to calculate the **conditional probability** of "A given B".

Bayes' theorem says: $P(A|B) = \dfrac{P(B|A)P(A)}{P(B)}$

This result can be used to test **hypotheses** in experiments, by calculating the probability of the hypothesis being true (A) given the **data** observed (B). It is used throughout science and medicine, and in finance and social sciences, to understand relationships between dependent events.

The results of a medical test must be considered in the context of the test's accuracy - a positive result might not mean you have a disease

	10,000 people (total)	Positive test	Negative test
Has disease (5%)	500 people	495	5
No disease (95%)	9500 people	95	9405
		19.19% (proportion of positive tests who do not have the disease)	0.05% (proportion of negative tests who have the disease)

82

Sampling

Statistical surveys and questionnaires are a great way to find out about the opinions and characteristics of a population. In the case of a census or election, every member of a population is polled and their **data** is collected for analysis. But it's not always practical to ask everyone – in which case, we need to choose a sample to survey.

A population in this context could be a group of people, animals, experimental samples, or even vehicles or equipment. Sampling allows us to use fewer resources when collecting data, then apply statistical tools to generalise these results.

WHY IT MATTERS
Carefully selected samples are necessary to reduce **sampling bias** – it's easy to find out what people who love filling in surveys think, but you need to find out what other people think as well!

KEY THINKERS
Pierre-Simon Laplace
(1749–1827)
Alexander Chuprov
(1841–1908)

WHAT COMES NEXT
Once data has been collected from a sample, statistical methods can be used to generalise the results to the whole population

"The purely random sample is the only kind that can be examined with confidence by means of statistical theory, but there is one thing wrong with it. It is so difficult and expensive to obtain for many uses that sheer cost eliminates it."

Darrell Huff, in *How to Lie with Statistics*

Sampling involves choosing a subset
of a population to test, in order to draw
conclusions about the properties of the whole
population. In order for conclusions about the sample to
be applicable to the larger set, the sample must be carefully
chosen to be representative of the whole.
Various sampling methods exist, each with different benefits.
Participants can be picked **randomly**, or the entire population
sorted in order and individuals picked at regular
intervals from the list. Stratified sampling involves
grouping participants into demographics,
then taking samples from each
group proportional to the
size of that group within
the population.

83

Correlation

Statistics allows us to describe trends and relationships in the **data** we observe. Sometimes, observed relationships are caused by underlying connections between **variables** – for example, if we observe that people who eat a lot of sugary foods have a high incidence of tooth decay, we might imagine that one is causing the other.

But there is a big difference between two variables appearing to be connected, and **proof** that increasing or decreasing one actually causes changes in the other. We use the term correlation to describe two variables which change together – but not necessarily because they're connected. There may be a third variable – a confounding factor – affecting both.

As mathematicians are fond of saying, correlation does not imply causation.

Positive Correlation

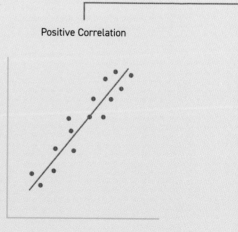

Given two **variables**, a correlation is a relationship between their values – for example, if one increases when the other increases (a positive correlation), or if one decreases when the other increases (a negative correlation). Correlations describe linear relationships – where the variables change together at a fixed rate. Correlated **data** points will appear to cluster along a line. We can measure the degree to which two variables are correlated using a correlation coefficient, which will be bigger if they are more strongly correlated. If their values follow one another closely, then one can be used to make predictions about the other.

Negative Correlation

No Correlation

84

Normal distribution

WHY IT MATTERS
The normal distribution
describes many
different types of data
sets. It emerges
unexpectedly in many
different situations
when handling data, so
it can be used to
assess how likely or
unlikely some outcome
is across many
contexts

KEY THINKERS
Carl Freidrich Gauss
(1777–1855)
Pierre-Simon Laplace
(1749–1827)

WHAT COMES NEXT
Once we know or guess
a situation is normally
distributed, we can
judge whether a
particular outcome is
unusual or unlikely
enough to be significant
evidence in a
hypothesis test

SEE ALSO
Probability, p.141
Statistics, p.142
Binomial
distribution, p.154

Plotting **data** on a graph lets us see its shape or distribution. Many different sets of data have a similar shape: peaking around some central value, and tailing off as you go away from this value. This shape can be modelled by a curve that has several names: a Gaussian distribution (after Carl Freidrich Gauss who first described it), a bell curve (although it's not the only distribution with a bell shape), or a normal distribution.

It's said that the name "normal" originally referred to the equations that Gauss used to describe the curve – these used properties of **perpendicularity** to derive the shape (*normal* in a technical sense means "at right angles"). This accidental name for the distribution has stuck because of how "normal" – in an everyday sense – this type of distribution is. Karl Pearson commented that a disadvantage of this is that it might give the (false) impression that other shapes of distribution are somehow abnormal.

That said, the **central limit theorem** is based on the idea that almost *any* **data** set, if you take samples from it, tends to lead to a normal distribution.

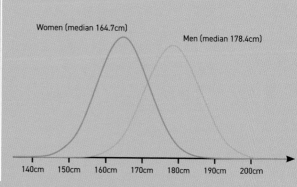

Women (median 164.7cm)

Men (median 178.4cm)

| 140cm | 150cm | 160cm | 170cm | 180cm | 190cm | 200cm |

In
100
words

The normal distribution models the common distribution of many real **data** sets, particularly when they represent physical properties with some deviation from an **average**. The curve is symmetrical: normal distributions have the same **mean** and **median** (loosely, the shape is not **skewed**). By adjusting the parameters of the model, you can change the curve's shape to approximate any chosen data set.

The central limit theorem states that in many situations, the distribution of the averages of different samples from the same data set will be a normal distribution – even if the original data was distributed in a completely different way.

Many data sets are well approximated by a normal distribution curve. The left graph shows distributions of people's heights from around the world; the right graph shows distribution of child birth weights (in grams) from the UK in 2021

1000g 1500g 2000g 2500g 3000g 3500g 4000g 4500g 5000g 5500g 6000g

85

Binomial distribution

WHY IT MATTERS
We can use binomial distributions to judge how unlikely an outcome is and whether it is strong enough evidence to suggest that the model is wrong (see page 158)

KEY THINKERS
Jacob Bernoulli (1655–1705)

WHAT COMES NEXT
Other discrete distributions describe different situations where we can count data; for example, the Poisson distribution describes situations like the number of people joining a telephone queue in a call centre over a certain period of time

SEE ALSO
Probability, p.141
Statistics, p.142
Normal distribution, p.152
Significance, p.158

Imagine you are asked ten hard multiple-choice questions, each with five choices for the answer. How many are you likely to get right just by guessing?

Situations like this, where you can count the number of **successes** from some **trials** and each success has a fixed probability, give distributions of outcomes called binomial distributions. For our quiz, the most likely number of correct answers is two, but other outcomes are possible. The bar chart shows the shape of the distribution of the possible outcomes, with the height of each bar representing the probability of that outcome. Any number of correct answers from zero to four is quite likely, but anything higher than that is very unlikely.

A variable which gives this distribution is called a binomial random variable. To calculate the probability of an outcome, such as the probability of four correct answers, you need to know the number of trials, n, and the probability of success for each trial, p. In our example, $n = 10$ and $p = 1/5$.

Binomial random variables are examples of discrete **random variables** (as opposed to continuous ones, like a normal random variable) because they only give probabilities for distinct outcomes rather than a continuous range.

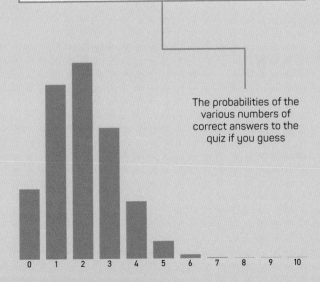

In 100 words

A binomial distribution is the distribution of possible outcomes when you count the **successes** from *n* **independent trials**, with a probability of **success** equal to p each time.

The binomial distribution is used to model a variety of situations: for example, the number of correct guesses in a multiple choice quiz, the number of defective items produced on a factory assembly line, or the number of people who might experience severe side effects for some medication.

If *n* is large enough, and p is not too small or large, the binomial distribution can be well approximated by a normal distribution.

The probabilities of the various numbers of correct answers to the quiz if you guess

0 1 2 3 4 5 6 7 8 9 10

155

86

Markov chains

WHY IT MATTERS
Markov chains can be used to model processes in physics, chemistry, and finance, as well as for signal processing and speech recognition

KEY THINKERS
Andrey Markov (1856–1922)
Tanja van Aardenne-Ehrenfest (1905–1984)
Andrey Kolmogorov (1903–1987)
Eugene Dynkin (1924–2014)

WHAT COMES NEXT
Language models used for predictive text are often built on Markov chains, by studying how often a particular word is followed by certain others and using it to make suggestions

SEE ALSO
Dynamical systems, p.133
Probability, p.141
Statistics, p.142

Markov chains allow us to model situations where probability is involved but the future possibilities depend on the current position – for example, in a board game where a dice is rolled to determine how many squares to move, the consequence of rolling a four will depend on where the player's piece currently is.

Almost any ongoing process can be modelled using a Markov chain, although the more complicated the situation, the more information you need to store as the current state in order to be able to predict future probabilities.

"The genie cannot be put back into the bottle. The Bayesian machine, together with Monte Carlo Markov Chains, is arguably the most powerful mechanism ever created for processing data and knowledge."

James O Berger, statistician

A Markov chain is a model used to describe probabilistic sequences of events, in which the probability of each possible future state depends only on the current state of the system. Markov chains allow us to model complex real-world systems that involve probability, such as **queueing simulations** or **Brownian motion**.

Markov chains can be discrete, so that each step happens separately, or continuous, where the situation is constantly being updated. The memoryless property of such models means that previous states are irrelevant, but we can still use the models to make statistical predictions about the likelihood of different future outcomes.

87

Significance

How unusual is too unusual? Is a particular outcome significant? These questions do not have precise right or wrong answers, but are often important.

If we can model a situation with a **probability distribution**, then we can calculate a probability of the particular outcome occurring. If the outcome is extremely unlikely to have happened with this distribution, then we might suspect that there is a different explanation and the model distribution we used is wrong.

Exactly how unlikely the result has to be before we start to suspect that the chosen distribution is wrong is called the significance. Commonly used significance values are 5% or lower, and should be agreed *before any results are known*.

Imagine again the quiz discussed for binomial distribution. If someone scored six or more, would you believe that they were doing something better than just guessing? In this case – a binomial distribution with $n = 10$ and $p = 0.2$ models – someone guessing, and getting a score of six or higher has a probability of less than 1%.

If the previously agreed significance level is higher than 1%, then this outcome might be agreed to provide evidence to suggest that the quiz-taker is not just guessing.

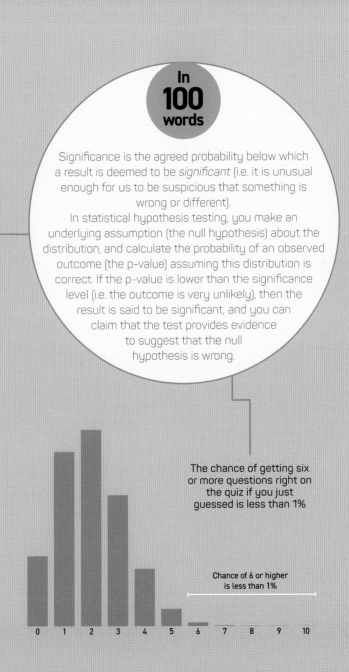

Significance is the agreed probability below which a result is deemed to be *significant* (i.e. it is unusual enough for us to be suspicious that something is wrong or different).

In statistical hypothesis testing, you make an underlying assumption (the null hypothesis) about the distribution, and calculate the probability of an observed outcome (the p-value) assuming this distribution is correct. If the p-value is lower than the significance level (i.e. the outcome is very unlikely), then the result is said to be significant, and you can claim that the test provides evidence to suggest that the null hypothesis is wrong.

The chance of getting six or more questions right on the quiz if you just guessed is less than 1%

Chance of 6 or higher is less than 1%

0   1   2   3   4   5   6   7   8   9   10

88

Game theory

WHY IT MATTERS
Nobel prizes in economics are often awarded to game theorists – a zero-sum game is a good model for economics, where resources are shared. Game theory also has applications to geopolitics, evolution, and sociology

KEY THINKERS
John Forbes Nash
(1928–2015)
John von Neumann
(1903–1957)

WHAT COMES NEXT
Combinatorial games are a particular class of games which can be mathematically analysed

SEE ALSO
Combinatorial games, p.162

Strategic decisions people make – either when playing simple games, or in high-level strategic thinking in economics or politics – can be modelled mathematically by considering the options available to each player, and the losses and gains that would result from a decision.

Game theory formalises this, by considering the properties of the game being played. Is it **cooperative** or **competitive**? Do the players take turns, or are moves played simultaneously? Is all information available to all players (if so, the game has perfect information), or is some kept secret? Game theory allows mathematically optimal strategies to be devised, which can be applied to real-world situations.

Game theory is the study of strategy
and decision-making in theoretical scenarios
involving completely rational players, which allows
us to model real-world situations including economics
and political strategy.
Games involve each player gaining or losing by some metric;
a zero-sum game is one in which the total amount of score
available is fixed, and one player's loss results in another's
gain. Games can be symmetric – both players have
the same choices and moves – or asymmetric.
An important concept is the idea of a
Nash equilibrium – a situation
where neither player can
improve their situation or
payoff by changing
their strategy.

"An arguing couple spiralling into
negativity and teetering on the brink
of divorce is actually mathematically
equivalent to the beginning of a
nuclear war."

Hannah Fry, mathematician

Combinatorial games

WHY IT MATTERS
Techniques developed in the study of simple game systems can be applied to search algorithms and to problems in planning and scheduling

KEY THINKERS
Charles Bouton
(1869–1922)
Elwyn Berlekamp
(1940–2019)
John Horton Conway
(1937–2020)
Richard K. Guy
(1916–2020)
Claude Shannon
(1916–2001)

WHAT COMES NEXT
We can apply knowledge of one game to any other game which is structurally identical. Any impartial game – where both players can make the same moves – is equivalent to some form of nim

SEE ALSO
Combinatorics, p.41
Game theory, p.160

In the real world, games can be messy and complicated. Some involve probability through dice rolls or card shuffles, and some involve figuring out whether something went over a line, who shouted "snap" first, or whether your opponent is bluffing. But there are some, called combinatorial games, which we can reduce to a simple mathematical set of ideas.

Combinatorial games include chess and noughts and crosses, as well as variations on existing games – like infinite chess – and one-player games or puzzles like Minesweeper or sudoku. One famous example is nim – a simple game believed to originate in ancient China, which involves taking turns to remove stones from piles. The winner can be the one who takes the last stone, or the one who forces their opponent to do so, and it can be played with different rules for the number of stones you can take in one turn, and different numbers and sizes of piles.

When analysing all possible moves or game states, the number grows as you progress through the game, and techniques from combinatorics can be used to count the possibilities. For example, chess has been shown to have at least 10^{120} possible game states.

A combinatorial game is a turn-based game in which all players have perfect information – everyone is aware of the full game state at all times. When a player takes a turn, they modify the game state in a way that is consistent with the rules of the game.

Such games can be studied logically using tools like game trees or by defining game state notation, and in smaller cases, optimal strategies can be defined. For larger and more complex games, like chess or the ancient Chinese game Go, there are many possible game states, and analysis requires huge computing power.

90

Zero-player games

To play a game, you might imagine you need two players – or at the very least, one person playing a game of solitaire. But it is possible to define a zero-player game: a form of combinatorial game in which one can observe the progression of game states, but have no influence over the outcome.

We can use the term to describe a game like snakes and ladders: a dice roll determines each move and the players don't make any decisions. It's also used to describe games where two programmed AIs play against each other.

The term is also applied to "games" which are logically determined by rules, proceeding from a starting position chosen by a human. Such games are often played on grids of cells and are called cellular automata.

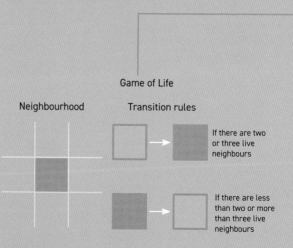

Game of Life

Neighbourhood   Transition rules

If there are two or three live neighbours

If there are less than two or more than three live neighbours

A zero-player game is one that follows predetermined rules from some initial state. A cellular automaton is a zero-player game based on a grid of cells, where the colours of cells change at each point in time according to certain rules, and the system evolves gradually over time.

The Game of Life is one example, where cells "live" or "die" (turn black or white) at each time step depending on how many of their neighbours are black. Humans can try to find initial configurations for the game which lead to interesting patterns – formations which persist or propagate across the grid.

There are many known configurations whose behaviour under the Game of Life rules has been studied – some are stable, while others flip between two states and others "die out"

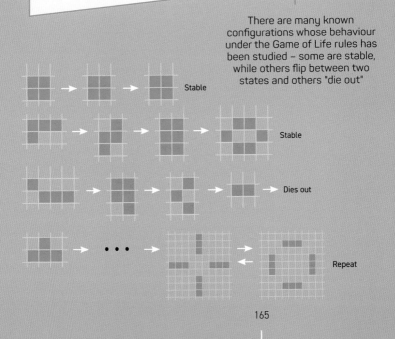

Stable

Stable

Dies out

Repeat

Logic and Proof

How do we know if a theory is correct? In many areas of science we rely on statistical techniques to provide evidence – and although it may mount up, we can never be completely certain. Mathematics, however, requires a rigorous, step-by-step proof of an idea before it is accepted.

The tools for proving mathematical ideas have developed over time, and every new idea proved provides another tool that can be used to prove further ideas. Well-defined logical systems such as Boolean logic have allowed us to develop computing systems, which has led to computers being used to prove ideas that were previously beyond reach. And areas such as set theory and category theory allow us to understand how different areas of mathematics are related and use tools from one area to prove ideas in another.

Even with these tools, some ideas stubbornly resist being proved, despite mathematicians working on them for hundreds of years. In fact, in the mid-1900s Kurt Gödel showed that any mathematical system will always have statements that are impossible to prove. Despite these difficulties, mathematical proof remains the central method by which any new mathematics is confirmed and accepted.

91

Mathematics is built on **proof**. Beyond a few basic **axioms** – fundamental statements that are agreed to be true – each further idea must be proved logically. Traditionally, a proof is a written argument that works through, step by step, the mathematics required to prove an idea. Whether this takes just a few lines, or hundreds of pages of text and equations, it allows other mathematicians to read and verify the proof. Once a **theorem** has been proved it becomes part of a toolbox that can be used to prove other **conjectures** or ideas, thus providing a solid basis for further mathematics.

Proof

You can think of mathematics as a structure built from **theorems, conjectures**, and **hypotheses**. Proof is the cement that holds them in place. You can build on top of a block that hasn't yet been cemented in, but if that block falls out then everything built on top of it will come tumbling down. Similarly, you could use an unproven conjecture to prove a new idea, but your proof would be conditional on the original conjecture being true (see page 56).

Advances in computing have led to new methods of proof. Sometimes it's possible to use a "brute force" approach, by trying every possible case to see if it's true. If there are millions of possible cases, this is far too time-consuming to do by hand, but a computer program can work through them all much faster. A disadvantage of this approach is that it doesn't tell us *why* a theorem is true. A computer proof might be followed up years later by a more satisfactory logical proof that offers more insight into the problem. And if there are an infinite number of possible cases, a computer will never be able to check them all.

WHY IT MATTERS
Proof is the cement that holds all of mathematics together

KEY MOMENTS
The four colour theorem (see page 128) was the first major theorem to be proved by computer (1976)

WHAT COMES NEXT
Proof assistant software is the next step in computerised proofs. Rather than checking each case, this software can be used to define and verify complex abstract concepts

SEE ALSO
Four colour theorem, p.128
Axioms, p.179
Gödel's incompleteness theorem, p.180
Millennium Prize Problems, p.181

92

Boolean logic

When George Boole introduced Boolean algebra in 1847, it was a purely mathematical concept. It wasn't until the 1930s that Claude Shannon developed it into the Boolean logic that led to the modern information age. Shannon first applied it to digital switching circuits, where the on/off position of each switch determined its truth value, and then by developing a system to encode any piece of text using 1s and 0s, which correspond to TRUE or FALSE. These basic concepts underpin almost every piece of computer hardware, from the Ferranti Mark 1, the first commercially available computer, to modern smartphones.

> "I am now about to set seriously to work upon preparing for the press an account of my theory of Logic and Probabilities which in its present state I look upon as the most valuable if not the only valuable contribution that I have made or am likely to make to Science and the thing by which I would desire if at all to be remembered hereafter."
>
> **George Boole**, mathematician

Boolean logic assigns one of two values to a statement; for example, the value of the statement "this switch is on" could be TRUE or FALSE. Its power comes from the **operators** AND, OR, and NOT, also called logic gates. The NOT operator gives the opposite value to the statement entered into it. Combining two statements with the AND operator gives a value of TRUE if both statements are true, and FALSE otherwise. The OR operator gives a value of TRUE if one or both statements is true. Combinations of operators can be used to perform calculations with binary numbers.

Propositional logic

Proof, p.167
Boolean logic, p.168
Axioms, p.179

Propositional logic builds on the ideas of Boolean logic – assigning TRUE or FALSE values to statements – to create more complex sentences whose truth can be determined.

For example, the **proposition** $A \lor B$ (meaning "A or B") is true if whenever one of A or B is false, the other is true. This might apply to statements like "x is an odd number" and "$x + 2$ is an even number".

We can employ logical tools like *modus ponens*: if we know A is true, and that $A \to B$ is true ("A implies B"), then we can infer B is true. For example, if A is "x is a dog" and B is "x chews bones", then $A \to B$ is equivalent to the statement "All dogs chew bones". This statement can be true regardless of whether or not A and B are true – it just means that if A is true, we know B is as well. This implication is one-way: if we know something chews bones, that doesn't mean it's a dog – it might be a wolf!

Formalising logical statements means we can clearly lay out the logic behind **proofs** – and use computer proof systems.

> "Mathematics is the study of anything that obeys the rules of logic, using the rules of logic."
>
> **Eugenia Cheng**, mathematician

Propositional logic, sometimes called propositional calculus, involves combining logical **propositions**, denoted by capital letters, using the Boolean **operators** AND (denoted ∧) and OR (∨), along with NOT (¬) and an implication arrow (→) for "if... then". The truth values of statements can either be TRUE or FALSE, and propositions define the relationships between them. For example, $A \to B$ means "if A is true, then B is true", and whether or not the proposition itself is true depends on A and B. The statement $A \lor \neg A$ (meaning "A or not A") is a tautology – it will always be true.

"The fact that all Mathematics is Symbolic Logic is one of the greatest discoveries of our age; and when this fact has been established, the remainder of the principles of mathematics consists in the analysis of Symbolic Logic itself."

Bertrand Russell, mathematician and philosopher

94

Venn diagrams

WHY IT MATTERS
Venn diagrams allow us to visualise the different ways in which a collection of sets can be related

KEY THINKERS
John Venn (1834–1923)
Khalegh Mamakani (PhD 2013)
Frank Ruskey (PhD 1978)

WHAT COMES NEXT
Duals (see page 86) of Venn diagrams are of interest in studying higher-dimensional geometry

SEE ALSO
Set theory, p.178

Venn diagrams show relationships between **sets** – these can be sets of numbers, or more abstract concepts. A typical Venn diagram uses two overlapping circles to represent two sets and their intersection. A two-set Venn diagram has four distinct areas (including the area outside both sets), and a three-set diagram has eight areas.

Using only circles, the maximum number of overlapping sets you can represent is three. For higher numbers of sets, you need to be more creative with the shapes you use, and the results get more complex – and more beautiful – the higher you go. In 2012, mathematicians Khalegh Mamakani and Frank Ruskey created a symmetric Venn diagram for 11 sets – it has 2,048 areas, representing every possible combination of sets.

Venn diagrams have become a popular form of meme, used to make jokes or political points about the overlap (or lack of overlap) between different cultures, ideologies or, in more light-hearted cases, food groups.

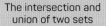

The intersection and union of two sets

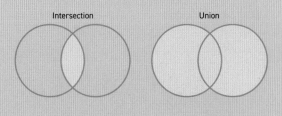

Intersection   Union

Venn diagrams are made of two or more regions with overlapping areas. Each region represents a **subset** of a set, and the overlap between two regions represents the intersection – the **elements** common to both sets. The union of two sets means the elements that are in one or the other, or both, and is represented by the total area of both regions, including the overlap. A Venn diagram shows all possible combinations of subsets, even if they contain no elements. An Euler diagram is similar, but only shows nonempty regions – so two sets with no common elements will not overlap.

Possible Venn diagram constructions for 5 sets (32 areas) and 7 sets (128 areas)

95

Logical paradoxes

WHY IT MATTERS
Logical paradoxes expose potential problems and inconsistencies in mathematics

KEY THINKERS
Zeno of Elea (c. 495BCE)
Betrand Russell (1872–1970)
Ernst Zermelo (1871–1953)
Alfred Tarski (1901–1983)
Kurt Gödel (1906–1978)

WHAT COMES NEXT
Fuzzy logic is a logical system that allows the truth value of a statement to be any value between 0 and 1. Using fuzzy logic, the liar paradox is no longer a paradox – it has a truth value of 0.5

A paradox is a self-contradictory statement or situation. Paradoxes have been used by mathematicians and philosophers since ancient times to illustrate problems in arguments and to stimulate thought and discussion, and they have often led to progress in mathematics.

In 1901, Bertrand Russell published a paradox in **set** theory: the set of all sets that are not members of themselves. Is this set a member of itself? If it is, then by definition it cannot be. But if it is not, then by definition it must be. The paradox showed that there were problems in the way set theory was defined, and it led to the development of first-order logic and a more rigorously defined notion of set theory, known as Zermelo–Fraenkel set theory.

A liar paradox is a self-referencing statement that is neither true nor false, such as "this statement is false". If the statement is false, then it must be true. Conversely, if the statement is true, then it must be false. Gödel used a variation of the liar paradox in his **proof** of the incompleteness theorem (see page 180).

A paradox arises if a statement is
both true and false at the same time, and
a paradoxical statement can often indicate a
problem or inconsistency in the system it is written in.
The sentence "this statement is false" demonstrates the
beauty, and inherent difficulty, of natural language – we don't
expect every statement to be either true or false, and
the truth value of a statement often depends on how
it is interpreted.
In mathematics, however, we want truth to
be rigorously defined and not subject
to interpretation. A system that
leads to a paradox is not
a well-defined
system.

96

Category theory

WHY IT MATTERS
Category theory
allows us to unify
different areas of
mathematics and
better understand how
they are related to
each other

KEY THINKERS
Emmy Noether
(1882–1935)
Samuel Eilenberg
(1913–1998)
Saunders Mac Lane
(1909–2005)
Emily Riehl (c. 1984–)
Eugenia Cheng (1976–)

WHAT COMES NEXT
Infinity category theory
is the study of
higher-dimensional
category theory

SEE ALSO
Group theory, p.76
Topology, p.90
Set theory, p.178

Category theory is often called the mathematics of mathematics. It builds on ideas from **set** theory, group theory, and topology to give a systematic way to study all types of mathematical structures, and to understand what is the same about them and what is different. In doing so, it promotes a deeper understanding of these structures.

Category theory's flexibility means it can be used to zoom in on tiny details of a particular structure, or zoom out to give a bird's-eye view. Where topology can be seen as an **abstraction** of geometry, and set theory as an abstraction of algebra, category theory can be seen as an abstraction of mathematics itself. By using the idea of equivalence, where two structures are considered the same in a defined way, it allows ideas to cross boundaries between areas of mathematics that were previously seen as completely unrelated.

> "Category theory was much more compelling to me and I loved and understood it in its own right, whereupon it helped me to understand all those other parts of pure math that I had never really understood before."
>
> **Eugenia Cheng**, in *The Joy of Abstraction*

Category theory analyses mathematical objects
based on their relationships to other similar objects.
A category consists of objects and the relationships
(called morphisms) between them. A category must
have an **identity** (a morphism from an object to itself)
and any two **morphisms** must combine to form a
composite morphism that is also in the category.
On a small scale, the factors of 24 form the objects
of a category, with the relationship "is a factor of".
On a larger scale is the category whose
objects are all possible groups and whose
relationships are maps between groups
(called group homomorphisms).

97 Set theory

Set theory formalises the idea of considering a collection of objects as a single entity. The theory of sets underlies much of mathematics: functions (see page 54) operate between sets, and counting can be considered as comparing the size of a set of objects to a set of numbers of the same size.

Much of the development of mathematical logic has been based on set theory, and mathematicians considering sizes of infinity (see page 27) were thinking about infinite sets. The notions of union and intersection roughly correspond to the OR and AND **operators** in logic (see page 168 and 170).

WHY IT MATTERS
Many mathematical concepts can be defined precisely using only concepts from set theory – including graphs (see page 126), objects in topology (see page 90), and vector spaces (see page 110)

KEY THINKERS
Georg Cantor
(1845–1918)
Richard Dedekind
(1831–1916)
Ernst Zermelo
(1871–1953)
Abraham Fraenkel
(1891–1965)

WHAT COMES NEXT
Adding structure to a set gives us more complex algebraic objects like **groups** (see page 76)

SEE ALSO
Integers and counting, p.9
Boolean logic, p.168
Propositional logic, p.170
Venn diagrams, p.172
Axioms, p.179

In
100
words

A set is a collection of objects (**elements**) and can be finite or infinite, contain other sets, or nothing at all. The empty set – the set with no elements – is denoted Ø, and sets are written as a list in braces: $\{a, b, c\}$. Sets have no defined ordering, and each element can occur only once. Two sets containing exactly the same elements are considered equal. We can define operations for combining sets, including union – all elements of both sets – and intersection – the elements found in both sets. A set which is entirely contained in another is called a **subset**.

An **axiom** is a basic premise on which further arguments are made. An axiom isn't proved; rather, it is explicitly stated as something that is assumed to be true in order to be able to make progress and draw further conclusions. For example, one of the basic axioms of arithmetic is that if $a = b$, then $b = a$. This might seem so obvious that it should never need to be said, but it sets out formally the meaning of equality and thus ensures that everybody working with arithmetic means the same thing when they say that one number is equal to another.

WHY IT MATTERS
Axioms are the foundational blocks upon which mathematics is built

KEY THINKERS
Euclid (c. 325–265 BCE)
Giuseppe Peano (1858–1932)

WHAT COMES NEXT
Once axioms have been defined, they can be used to prove many more complicated ideas

SEE ALSO
Euclid's
Elements, p.100
Proof, p.167

Axioms

An axiom is a basic fact that everybody agrees to be true (but possibly only in a particular context). Stating axioms clearly is a good idea in many situations. If the participants in a debate don't agree on some basic beliefs at the outset, it's easy for them to (accidentally or intentionally) misunderstand one another. Mathematicians aim to avoid these misunderstandings by defining everything clearly, no matter how obvious such definitions might seem.

Different areas of mathematics use different sets of axioms. For example, Euclid defined five axioms (or **postulates**) for geometry, and Peano defined nine axioms for natural numbers.

99 Gödel's incompleteness theorem

Gödel's incompleteness theorem proved that in any axiom-based system of mathematics there are true statements that cannot be proved from the **axioms**. This came as a shock to many mathematicians, who believed in the completeness and power of their subject. Before Gödel, it was widely assumed that any mathematical statement could be either proved or disproved – though some might be very difficult and possibly take hundreds of years, that was merely due to limitations of technique and the human brain, not a limit of mathematics itself. Gödel's theorem placed boundaries on the ability of mathematics to solve any problem.

In **100** words

Mathematics is based on axioms – basic facts that are agreed to be true and from which everything else is proved. A set of axioms should be consistent – i.e. they should not lead to **contradictions**. Gödel used mathematical methods to study mathematics by devising a method to encode every mathematical statement as a number. With this method, he proved that any consistent set of axioms must also be incomplete: there will be true statements that cannot be proved using only those axioms. Gödel originally stated his **theorem** in terms of number theory, but it can be extended to other axiomatic systems.

The Millennium Prize Problems are a set of seven open problems announced by the **Clay Institute** in 2000 as the most important unanswered questions in mathematics. Each has a $1m prize, and they range across all areas of mathematics, from number theory (the Riemann hypothesis) and fluid dynamics (the **Navier–Stokes equations**) to computability (P vs NP) and questions from geometry and topology. Each problem represents a huge breakthrough in its area. Of the seven original problems, only one has been solved – the **Poincaré conjecture**, concerning four-dimensional spheres, was proved in 2010 using a method involving partial differential equations.

WHY IT MATTERS
Prizes like the Millennium Prize Problems motivate discovery and show the public that mathematics is a rich field with many open questions

KEY THINKERS
David Hilbert (1862–1943)
Henri Poincaré (1854–1912)
Grigori Perelman (1966–)
Andrew Wiles (1953–)

WHAT COMES NEXT
Solving more of these problems will unlock new areas of mathematics and provide answers to long-standing questions

SEE ALSO
The Riemann hypothesis, p.56
Differential equations, p.64
Complexity and sorting, p.122
P vs NP, p.124

Millennium Prize Problems

Mathematicians have many motivations for their work – the prestige of research publications, the joy of solving a puzzle, and the beauty found in discovering how the universe fits together.

But sometimes, people need a nudge towards the right kinds of problems. In 1900, mathematician David Hilbert picked a set of 23 unanswered questions from mathematics, and announced them as the Hilbert Problems, in order to motivate work on particular questions. Many have been solved, some remain open and others have been proven unknowable.

But Hilbert's questions provided inspiration for much great work, and for other similar problem sets over the years – including the **Clay Institute's** Millennium Prize Problems.

Glossary

Abstraction: Abstraction is a representation of a structure that contains only the most important, or most relevant, properties.

Algebraic (number): An algebraic number is a root of a polynomial with rational coefficients – a value which makes the polynomial equal to zero.

Algebraic structure: A collection of objects with a defined operation to combine them - like a group, or a vector space - with a set of axioms that define how the structure behaves.

Angular coordinate system: An angular coordinate system uses angles from one or more axes along with distances to define position.

Assumptions: An explicitly stated idea that simplifies a problem. For example, when estimating the number of words on a page, you could assume that each line contains the same number of words.

Average: A measure that represents a typical value of a data set.

Axes: An axis (plural axes) is a number line that is used to define position.

Axiom: A statement used as a basic premise, explicitly stated and assumed to be true, on which further arguments are built.

Bernoulli numbers: The Bernoulli numbers are a sequence of numbers that arise in many areas of number theory and analysis.

Brownian motion: The random movement of particles in a liquid or gas.

Category: A set of objects, together with a set of relationships between the objects – a generalisation of many ideas in mathematics.

Central limit theorem: A statistical theorem based on the idea that sample means from almost any data set tend to become normally distributed with increasing sample size.

Clay Institute: The Clay Mathematics Institute is a global organisation that works to promote and further mathematical knowledge.

Coefficients: A number written in front of a variable (or power of a variable) in an expresson or equation.

Competitive: A competitive game is one in which the players work against each other; an advantage to one player is a disadvantage to another.

Complex (number): A complex number is the sum of a real number and an imaginary number.

Concave: A concave polygon is one that is not convex; it will have at least one corner that points inwards and at least one internal angle greater than 180°.

Conditional probability: The probability of one event given that another event has occurred.

Conjecture: A conjecture is a mathematical idea that has been formally stated, but not proved.

Continuous (function): A continuous function is one whose output behaves sensibly as you vary the input – when plotted on a graph, it gives a single smooth curve.

Contradiction: A contradiction occurs if a statement is inconsistent with either itself or with known facts. It can be used as a method of proof - by showing that a statement leads to a contradiction, you prove that it is false.

Convex: A convex polygon is one with no inward-pointing corners.

Cooperative: A cooperative game is one in which the players work together to achieve a common aim.

Countably infinite: A set is countably infinite if its elements can be paired one-to-one with the natural numbers.

Cyclic: A cyclic polygon is a polygon where one circle can be drawn that passes through all its vertices.

Data: Facts and measurements that are collected for analysis.

Decimal places: Decimal places are the digits that appear after the decimal point in a number written in base 10.

Decryption key: A decryption key allows you to decrypt a coded message; the form it takes depends on the method used to encrypt the message.

Derivative: A function that defines the rate of change of another function.

Dimensionless quantity: A number used in a formula which is not associated with a particular unit of measurement, like a coefficient or scale factor.

Diophantine equation: An equation to which only integer solutions are of interest.

Diverge: If a sequence or series does not converge to a specific value, it is said to diverge.

Element: An element is a member of a set.

Ellipse: An oval-shaped curve defined by all the points for which the sum of their distance from two fixed focus points is the same.

Elliptic curves: An elliptic curve is a curve defined by an equation of the form $y^2 = f(x)$.

Equilateral triangle: A triangle with all three sides having equal length, and so also three equal angles (60° each).

Exponentials: A function involving a number raised to a varying power, often using the number e as the base.

Expression: A mathematical sentence or phrase made of numbers and variables, joined by operations.

Factor: A factor, or divisor, is a number that divides into another exactly, leaving no remainder.

False negative: A false negative occurs when a test gives a negative result but should have given a positive one.

False positive: A false positive occurs when a test gives a positive result but should have given a negative one.

Finite: An object that is finite has a boundary, or a start and end point. A finite set has a limited number of elements (even if the number is not known).

First-order logic: First-order logic builds on propositional logic by adding in quantifiers like "there exists" and "for all".

Fundamental theorem of calculus: The fundamental theorem of calculus links the rate of change of a function (calculated using differentiation) with the area under its graph (calculated using integration).

Game state notation: Game state notation is a method of describing states that may occur during a game; the exact notation depends on the game, but it will often include the position of pieces and which player has the next move.

Game tree: A game tree is a graph whose vertices represent positions in a game and whose edges represent moves on that game.

Generalise: To generalise means to find a pattern or structure that allows you to apply a result for a specific case to many other cases.

Gradient: The slope or steepness of a line or curve, which measures the rate of change.

Group: A mathematical structure consisting of a set of objects which can be combined using a group operation, obeying certain rules.

Group operation: A way of combining elements of a group; for example, multiplication or addition.

Hyperbola: A curve consisting of two bow-shaped parts, defined by all the points for which the absolute value of the difference of their distances from two fixed focus points is the same.

Hyperbolic geometry: Hyperbolic geometry takes place on a hyperbolic surface, and points in the space get further apart as you move away from any given point.

Hypotheses: Mathematical ideas that haven't necessarily been proven to be true, and which one might use a statistical test to suggest whether they might be true or not.

Identities: (As an equation) An identity is an equation that is true for any value of the variables; for example, $x + x = 2x$ is true for any value of x.

Imaginary (number): A number that is some multiple of i, the square root of negative one.

Independent: If two events are independent, the outcome of one does not affect the outcome of the other.

Insertion sort: A sorting algorithm that works through a data set item by item, putting them in the correct order as it goes.

Integers: The whole numbers, including zero and negative whole numbers.

Integral: A function that defines the area under the graph of another function.

Inverse: When an object is combined with its inverse it will "cancel out" - under addition, the inverse of positive numbers are their negative equivalents, and applying a function then its inverse has no result.

Irrational (number): A number whose decimal expansion goes on forever with no repeat, and which can't be written as a fraction of two integers.

Iterate: To apply a function or algorithm repeatedly, using the output to create the new input, to observe long-term behaviour.

Knot theory: A branch of topology concerned with knots – arrangements of closed loops in three-dimensional space – which can be studied using topological techniques.

Mathematically similar: If one shape is an enlargement of another, then the two shapes are mathematically similar.

Mean: An average found by adding all the members of a data set and dividing the result by the number of members.

Median: An average found by putting all the members of a data set in ascending order and taking the middle value (or the mean of the middle two values if the number of members is even).

Merge sort: A sorting algorithm that works by sorting smaller data sets before joining them back together into a fully sorted data set.

Modelling cycle: The process of repeatedly proposing, solving, checking, and upgrading a mathematical model to optimise how it describes a real-world situation.

Modulus: A value defined as equivalent to 0 in modular arithmetic, creating a finite set of elements we can use to perform calculations by taking the remainder on division by the modulus.

Modular functions: A complex periodic function; they are important in complex analysis and number theory.

Monster group: The Monster Group, also known as the Friendly Giant, is a large but finite mathematical group which contains over 8×10^{53} elements.

Morphism: A relationship between two objects in a category.

Multivariate analysis: A type of statistical analysis involving more than one variable or measurement.

Navier-Stokes equations: The Navier-Stokes equations are a system of differential equations that describe the movement of a fluid; it is not currently known if a solution to the equations always exists.

Nonlinear dynamics: The study of dynamical systems involving nonlinear equations – ones for which the change of the output is not proportional to the change of the input, like polynomials of degree 2 or higher.

Nonlinear relationship: A relationship between two variables that does not form a straight line when graphed.

Number theory: The study of integers.

Objective function: A function describing a system of related quantities, where the goal is to find the input that gives the maximum (or minimum) possible output value.

Operator: In Boolean logic, an operator is a way of combining two statements to produce a third statment.

Optimisation problems: An optimisation problem seeks to find the best solution from all the possible solutions, for example the solution that gives the maximum value.

Origin: The origin of a coordinate plane is the point at the centre, denoted $(0,0)$ in two dimensions.

Parabola: A U-shaped curve defined by all the points with the same distance from a fixed line (the directrix) and a fixed point (the focus).

Perpendicular: Two lines or planes that are at right angles (90°) to each other are said to be perpendicular.

Plane: A flat surface that extends infinitely in all directions.

Poincaré Conjecture: Asks if the surface of a four-dimensional sphere is simply connected; that is, can any loop on the surface be contracted to a single point without breaking either the loop or the surface?

Postulate: Postulate is another word for axiom, used in Euclid's *Elements*; basic facts that are agreed to be true and from which everything else is proved.

Power: A power is the number of times a number is multiplied by itself within an expression, indicated by a superscript – for example, in a squared term, the power is 2.

Power of 2: A power of 2 is a number that results from multiplying 2 by itself a number of times; the nth power of 2 is 2^n.

Prefixes: A set of letters that go in front of a unit to indicate that the quantity is a specific number of times bigger or smaller than the basic unit.

Probability distribution: A function which assigns a probability to each different possible outcome, from a range which can be discrete or continuous.

Product: The value obtained by multiplying two or more numbers together.

Proposition: In logic, a proposition is a statement that can be determined as either true or false.

Proof: A proof is a logical argument that can be verified by other mathematicians and that shows, beyond doubt, that a statement is true.

Pseudo-random numbers: Numbers which appear to be statistically random, but have been produced by a deterministic process which could be repeated, starting from the same seed value, to produce the same result.

Quantum cryptography: A highly secure method used for encrypting data that takes advantage of properties of quantum mechanics.

Quantum mechanics: The theory of atoms and subatomic particles.

Qubit: The basic unit in quantum computing, equivalent to a binary bit in conventional computing.

Queueing simulations: A tool to model and analyse the properties of queues, in order to predict average lengths and waiting times.

Quicksort: A sorting algorithm that involves repeatedly dividing a data set into smaller data sets according to whether the items are larger or smaller than a chosen value.

Random variable: A function defined by the outcomes of random events.

Random(ly): In a random sample, any member of a population has an equal chance of being chosen. In a random event, any of the possible outcomes has an equal chance of occurring.

Rate of change: The rate of change of one variable with respect to another describes how the first variable changes as the second increases by one unit.

Ratio: A ratio shows how many times one number is contained within another.

Rational (number): A rational number is any number that can be written as a fraction of two integers.

Real number: The set of real numbers is made up of all the rational numbers along with all the irrational numbers.

Roots: The solution(s) to a polynomial expression – values which, when put in as the variable, give zero.

Sampling bias: Sampling bias occurs when the sample taken from a population is not representative of the population, meaning that the data collected and the conclusions drawn may not apply to the population as a whole.

Set: A set is a collection of objects (elements). It can be finite or infinite, and it can contain other sets, or contain nothing at all.

Simple (polygon): A simple polygon has no intersecting edges and no holes.

Skewed: In a skewed normal distribution, the mean and median are not the same, resulting in an asymmetric bell-shape where the highest point is not in the centre.

Special and general relativity: Theories about the relationship between space and time, published by Albert Enstein.

Spherical geometry: Spherical geometry takes place on a sphere, and the shortest line between two points is a section of the line defining the circumference of the sphere that passes through those points.

Spread: Measures of spread show how spread out a data set is; the range is the difference between the largest and smallest member, and the interquartile range is the range of the middle section of the data set.

Square root: The square root of a number is the value that, when multipled by itself, gives that number.

Subset: A set which is entirely contained in another set.

Success: In probability trials, a success means that the event being tested for has occurred.

Surd: The irrational square root of a non-square number, written using a square root symbol.

Theorem: A mathematical idea that has been proved.

Tractrix: A curve defined by one end of a straight line when the other end is pushed and then dragged along a horizontal line; it has a cusp point from which it curves towards, but never reaches, the horizontal line.

Transcendental: A transcendental number is a number that is not the solution of any polynomial with only rational coefficients.

Trial: In statistics, a trial is one instance of a repeated experiment.

Uncountably infinite: A set is uncountably infinite if its elements cannot be paired one-to-one with the natural numbers.

Variable: In statistics, a variable is a value that changes; an independent variable is one that is changed intentionally by the researchers; a dependent variable is one that is measured to see the effect of changing the independent variable.

Velocity: The speed of a moving object together with the direction – often written as a vector.

Weighted graph: A network of points and lines where each line is assigned a numerical "weight", which could represent travel time or cost, for example.

Index

About the authors

Sam Hartburn is a freelance mathematician, author, editor and proofreader based in Whitstable, Kent. She obtained a BSc in Mathematics from York University in 2000. She has written articles for publications including *Chalkdust* magazine, and has contributed to several books including *Short Cuts: Maths*. As an editor and proofreader she has worked on more than 300 books and articles, spanning the spectrum from primary school to undergraduate and from academic research to recreational maths. In her spare time she writes and performs songs about maths, some of which can be seen on YouTube, and occasionally bakes mathematical cakes.

Ben Sparks is a mathematician, musician, and public speaker who gives maths talks and workshops around the world, to students, teachers, and the general public. He also works part time with the Advanced Maths Support Programme (AMSP) based at the University of Bath, as a speaker, tutor and co-ordinator. He has co-authored several books of problem-solving resources for A-level Further Mathematics, and appeared in many YouTube videos on the *Numberphile* channel, presenting interesting ideas in mathematics and using interactive visualisations.

Katie Steckles is a mathematician based in Manchester, who delivers maths talks, workshops and events around the UK, and appears on YouTube, TV and radio to share interesting mathematical ideas. She has written and contributed to several popular maths books including *The Math of a Milkshake*, *Short Cuts: Maths* and *The Curious World of Scientific Symbols*. She also writes mathematical puzzles and articles, including a column in *New Scientist* and for several mathematical blogs. In 2016 she won the Joshua Phillips Award for Innovation in Science Engagement, and was the London Mathematical Society popular maths lecturer for 2018.

In association with the Science Museum

The Science Museum is part of the Science Museum Group, the world's leading group of science museums that share a world-class collection providing an enduring record of scientific, technological and medical achievements from across the globe. Over the last century the Science Museum, the home of human ingenuity, has grown in scale and scope, inspiring visitors with exhibitions covering topics as diverse as robots, code-breaking, cosmonauts and superbugs. www.sciencemuseum.org.uk.

The publisher would like to thank the following for their kind permission to reproduce their photographs:

(Key: a-above; b-below/bottom; c-centre; f-far; l-left; r-right; t-top)

1 Shutterstock.com: art_of_sun. **2-3 Shutterstock.com:** Beebox Designs. **4-5 Shutterstock.com:** Foryoui3. **8 Shutterstock.com:** james weston. **20 Alamy Stock Photo:** The History Collection. **24-25 Shutterstock.com:** Jan Machata. **27 Shutterstock.com:** Aha-Soft. **39 Shutterstock.com:** galacticus. **41 Shutterstock. com:** kobeza (b). **44 Shutterstock.com:** Marina Sun. **52-53 Shutterstock.com:** Patthana Nirangkul. **64-65 Shutterstock.com:** Fouad A. Saad. **66-67 Shutterstock.com:** Pyty. **68-69 Alamy Stock Photo:** Science History Images. **76-77 Shutterstock.com:** pollapats. **79 Shutterstock.com:** Sunspire. **88 Shutterstock.com:** graphixmania (cb); Maria.K (crb). **89 Shutterstock.com:** Alex Illi (b). **90-91 Shutterstock.com:** Maxim Gaigul. **92 Science Photo Library:** Monica Schroeder (crb). **92-93 Shutterstock.com:** art_of_sun. **93 Dreamstime. com:** Martin Green / Mrgreen (cb). **96-97 Shutterstock.com:** Distance0. **106-107 Shutterstock.com:** vector punch. **108 Shutterstock.com:** neuralsuperstudio (b). **111 Shutterstock.com:** Ilya Bolotov. **114-115 Shutterstock.com:** Nickolay Grigoriev. **119 Shutterstock.com:** Nickolay Grigoriev. **122-123 Shutterstock.com:** Picksell. **126-127 Shutterstock.com:** Rainer Lesniewski. **133 Shutterstock.com:** cybermagician. **136-137 Shutterstock.com:** Robert Adrian Hillman. **138-139 Shutterstock.com:** cybermagician. **140 Shutterstock.com:** Beebox Designs. **142-143 Shutterstock. com:** Alexandr III. **148-149 Shutterstock.com:** KostiantynL. **156-157 Shutterstock.com:** Milles Vector Studio. **162-163 Shutterstock.com:** Apostle. **166 Shutterstock.com:** Lina_Lisichka. **168-169 Shutterstock.com:** Foryoui3. **176-177 Shutterstock.com:** Barks

Cover images: *Front:* **123RF.com:** joingate br; **Dreamstime.com:** Hendraxaverius tr, Microvone cr, Ttretjak cl; *Back:* **Dreamstime.com:** Microvone cra, clb; *Spine:* **Dreamstime.com**

All other images © Dorling Kindersley

DK | Penguin Random House

DK LONDON
Editor Florence Ward
Art Editor Anna Formanek
Managing Editor Pete Jorgensen
Managing Art Editor Jo Connor
Production Editor Jennifer Murray
Production Controller Louise Minihane
Publishing Director Mark Searle

Written by David Sang
Designer Neal Cobourne
Jacket Designer Steven Marsden

DK would like to thank Caroline Orr for copyediting, Alisa Jordan Walker
and Caroline Curtis for proofreading, Vanessa Bird for indexing and Izzy
Merry for design assistance.

First American Edition, 2024
Published in the United States by DK Publishing
1745 Broadway, 20th Floor, New York, NY 10019

A catalog record for this book
is available from the Library of Congress.
ISBN 978-0-7440-8161-9

DK books are available at special discounts when purchased
in bulk for sales promotions, premiums, fund-raising, or educational use.
For details, contact: DK Publishing Special Markets,
1745 Broadway, 20th Floor, New York, NY 10019
SpecialSales@dk.com

Printed and bound in China

www.dk.com